Mapping the Nation

Esri Press
Redlands, California

Esri Press, 380 New York Street, Redlands, California 92373-8100
Copyright © 2016 Esri
All rights reserved.

Printed in the United States of America
19 18 17 16   1 2 3

Ask for Esri Press titles at your local bookstore or order by calling 800-447-9778, or shop online at esri.com/esripress. Outside the United States, contact your local Esri distributor or shop online at eurospanbookstore.com/esri.

Esri Press titles are distributed to the trade by the following:

In North America:
Ingram Publisher Services
Toll-free telephone: 800-648-3104
Toll-free fax: 800-838-1149
E-mail: customerservice@ingrampublisherservices.com

In the United Kingdom, Europe, Middle East and Africa, Asia, and Australia:
Eurospan Group
3 Henrietta Street
London WC2E 8LU
United Kingdom
Telephone: 44(0) 1767 604972
Fax: 44(0) 1767 601640
E-mail: eurospan@turpin-distribution.com

All images courtesy of Esri except as noted.

Content

Foreword

Introduction

Content

Fostering a Healthy Nation | 45

Supporting a Resilient Environment | 63

Optimizing National Mapping and Statistics | 77

Elevating Education | 93

Content

Empowering Humanitarian Efforts | 103

Foreword

The pace of change today is greater than any other time in our history. Faster computing, real-time data, mobile devices, and cloud services are changing the way modern governments operate. At the same time, national governments face challenges such as population growth and climate change, stalled transportation systems and economic swings. Through this evolution, however, geographic understanding continues to provide a solid foundation for governments to support smarter, safer, healthier communities.

Geographic information systems (GIS) support deep knowledge of issues that leads to thoughtful action. Strategic organizations use Web GIS to share foundational data and support workflows across disciplines like health, transportation, and safety. Web GIS also enables organizations to implement sleek, ready-to-use mobile GIS apps and open data sites that provide great value to staff and citizens, fostering transparency, accountability, and efficiency. Web GIS plays an important role in the knowledge sharing that builds smart communities.

In this book you'll discover real examples of government agencies using GIS to implement smarter practices. Use it as a guide to find ideas that resonate with you, that can fuel your next project. You'll see that unlike any other technology, Esri's ArcGIS platform gives governments a reliable, proven way to analyze data, visualize impacts, and communicate with staff and the public. Dive in and explore how ArcGIS can empower your organization to support smart communities that build smart nations and a better world.

Warm regards,

Jack Dangermond

Introduction

Building Smart Government with GIS

Across the globe, governments use GIS to move communities toward healthier, more resilient, and safer outcomes. From actions as simple as optimizing routes for field workers, to visualizing policy impacts and developing intelligent transportation systems, GIS supports governments at all levels—local, state, and national.

Mapping, charting, surveying, land administration, and statistical organizations supply the foundational GIS data such as census counts and imagery that is critical to national, state, and local governments. Today, government organizations increasingly use Web GIS portals to share this data in real-time along with maps, apps, and services. The abundance of real-time data available now informs dynamic maps that support workflows for health, transportation, and public safety to name a few.

Web GIS portals make data easy to visualize, analyze, and share, creating more efficient organizations. This trend is transforming systems of record into systems of engagement that foster safe, stable, scalable environments to serve data, maps, and services that people can easily discover and use.

At the senior levels of government, leaders use GIS to inform policy decisions on issues that shape national and global action. Exploring issues with a geographic understanding reveals connections, patterns, and impacts that would otherwise go unnoticed. Maps also give leaders powerful visual tools to communicate with the public. People intuitively understand maps and the knowledge shared in this way with the public fosters an informed, engaged citizenry.

Mapping the Nation: Building Smart Government with GIS illustrates how governments use GIS to enhance national security, public policy, health, resilience, national mapping and statistics, humanitarian efforts, infrastructure, and education. Discover real-world practices that are the building blocks of smart communities and smart nations. Ultimately, these practices together can create a National GIS for a more prosperous, sustainable, and healthy world.

ARCHITECTING
A SAFE NATION

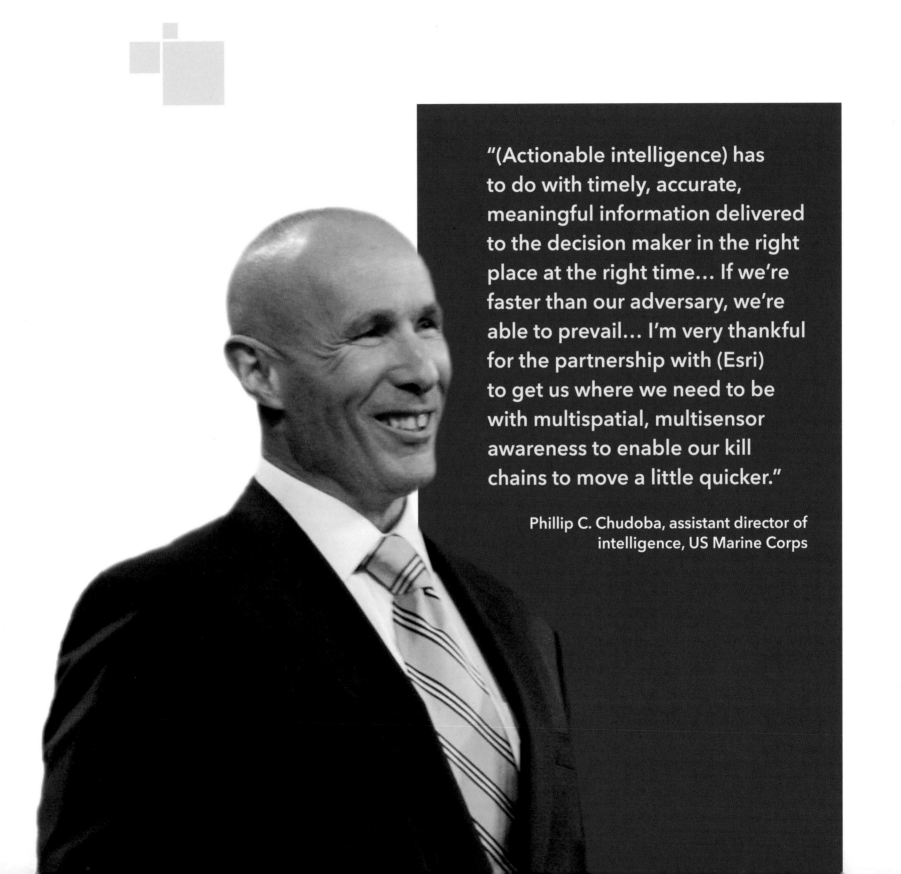

"(Actionable intelligence) has
to do with timely, accurate,
meaningful information delivered
to the decision maker in the right
place at the right time... If we're
faster than our adversary, we're
able to prevail... I'm very thankful
for the partnership with (Esri)
to get us where we need to be
with multispatial, multisensor
awareness to enable our kill
chains to move a little quicker."

Phillip C. Chudoba, assistant director of
intelligence, US Marine Corps

Architecting a Safe Nation

Ensuring public safety is the fundamental responsibility of many agencies and organizations at all levels of government. Senior leaders use GIS to combine data from many sources, analyze situations, predict outcomes, and align efforts to improve public safety. This approach creates awareness, understanding, and ultimately, smart communities.

At the strategic level, GIS informs policy makers and identifies risks through the use of maps. Online maps are a compelling way to present complex data in a visual, easy to understand way. By sharing information between agencies, GIS provides a platform for improved agency cooperation and effectiveness.

Operationally, web maps and operation dashboards provide commanders and first responders with critical real-time information necessary to protect public safety. Web GIS enables field operators to view and update maps from any mobile device.

Maps create a vital bridge between policy makers and the public. Stronger, safer communities result when citizens are engaged, informed, and empowered to respond. Whether at the national, state, or local level, officials turn to GIS to build smart communities that support safe nations.

Ball State University

Atmospheric Rivers and Heavy West Coast Precipitation Events

This is an educational poster concerning atmospheric rivers focusing on the west coast of the United States. Atmospheric rivers are relatively narrow regions in the atmosphere that transport water vapor outside of the tropics. The poster provides information concerning those regions that are particularly vulnerable to the effects of an intense atmospheric river. This student poster provides educational information to meteorology students concerning a synoptic (large-scale) weather pattern that affects the United States.

Atmospheric Rivers and Heavy West Coast Precipitation Events:
Typical Wintertime Synoptic Weather Pattern

Atmospheric rivers are relatively narrow, water-vapor rich regions of the atmosphere that are responsible for much of the horizontal transport of water vapor outside of the tropics. In the eastern Pacific, atmospheric rivers are the product of interactions between various high and low pressure systems that alter the position of the jet stream and effectively funnel near-tropical moisture towards the western United States. As this highly concentrated moisture reaches the coastal regions of the western United States, it is orographically forced up the many coastal and near coastal mountain ranges that span the length of the country. This orographic uplift helps to induce condensation as the moisture-laden air is forced to rise and cool. This process ultimately leads to the formation of precipitation, often in the form of either rain or snow. Atmospheric rivers can occur in many shapes, sizes, and durations but those with high amounts of water vapor, strong winds, and stall over vulnerable watersheds, can lead to extreme rainfall and flooding. However, not all atmospheric rivers are catastrophic. Many are weak and thus provide beneficial rain or snow, which is crucial to the rejuvination of water supplies across this region of the country.

Atmospheric River Facts and Info

On average, about 30-50% of the west coast's annual precipitation occurs in just a few atmospheric river events

A strong atmospheric river can transport an amount of water vapor roughly equivalent to 7.5-15 times the average flow of liquid water at the mouth of the Mississippi River

On average, atmospheric rivers are 400-600 km wide, but vary depending upon each event

Atmospheric rivers are a primary and important feature in the global water cycle

Most west coast atmospheric river events occur during the winter

A well-known example of a type of atmospheric river that can hit the west coast of the United States is the 'Pineapple Express'. The name is appropriate given their apparent ability to bring moisture from the tropics, near Hawaii, to the west coast

West Coast Effects

Flood Related FEMA Declarations
By County, 1953-Present

Total Number of Declarations
- 5 or Fewer
- 6-10
- 11-14
- 15 or More

Climate and FEMA Disaster Data
By State or City

Percentage of FEMA Disaster Declarations by Type
Washington, 1953-Present

Yearly Precipitation Distribution by Month
Olympia, Washington

Percentage of FEMA Disaster Declarations by Type
Oregon, 1953-Present

Yearly Precipitation Distribution by Month
Eugene, Oregon

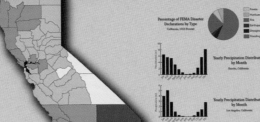

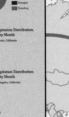
Percentage of FEMA Disaster Declarations by Type
California, 1953-Present

Yearly Precipitation Distribution by Month
Eureka, California

Yearly Precipitation Distribution by Month
Los Angeles, California

Source: U.S. Climate Data, Monthly Climate Summary, FEMA Disaster Declarations Summary

Source: Federal Emergency Management Agency, FEMA Disaster Declarations Summary, 1953-Present

Typical Atmospheric Progression
Prior to Heavy Precipitation Events

A Week or More Before Event

Strong Polar Jet Stream

Strong 'Blocking' High

Moisture Plume Over the Western Pacific

3-5 Days Before Event

'Blocking' High Weakens and Shifts Northwest

Weakened Split Jet Forms

Moisture Plume Extends Northeast

Heavy Rain Event

'Blocking' High Continues to Weaken

Strong Low Develops

Moisture Plume Extends Into the NW United States

Source: National Oceanic and Atmospheric Administration

Mitchel Pettit

US Air Force and Marstel-Day

The Joint Base Elmendorf-Richardson Mission

This map describes and depicts the installation's missions within the region to educate the stakeholders of Joint Base Elmendorf Richardson (JBER). Military installations and local communities are becoming increasingly intertwined physically, socially, and economically. As a result, there is a growing recognition within the Air Force to proactively coordinate with neighboring stakeholders to preserve both the military's operational capabilities with the community's quality of life. This map was included in the "Partners in One Community" document, a brochure intended to improve the public's awareness of JBER missions within the South Central Alaska region. This map will improve JBER's ability to find solutions with key community stakeholders and preserve both the military's missions as well as the quality of life for Alaskan citizens.

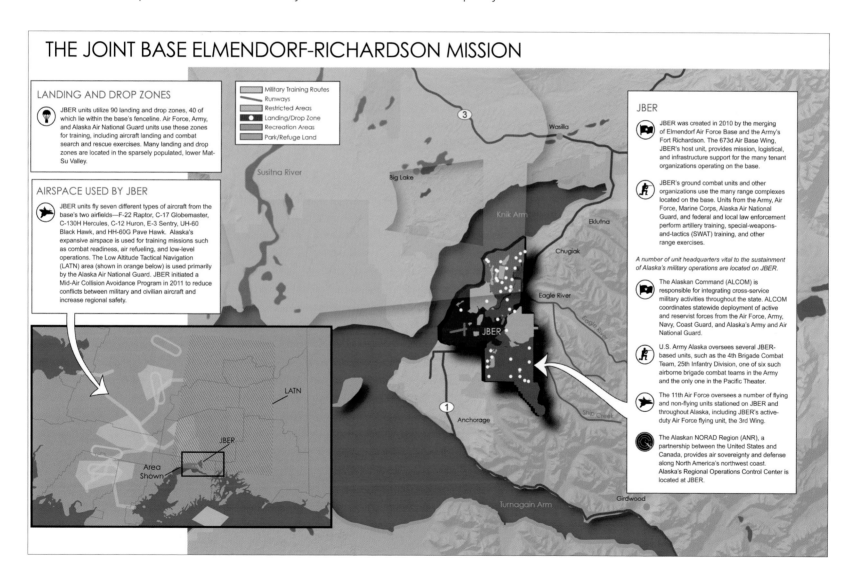

US Department of Agriculture (USDA) Forest Service

Kernel Density Wildland Fire Acres Burned Analysis

This map displays results of a kernel density analysis of wildland fire occurrence, by size, on the Francis Marion National Forest in South Carolina. The map addresses specific wildland fires with large growth characteristics and identifies historical patterns of large wildland fire growth which will aid fire managers in hazardous fuels treatment in these areas.

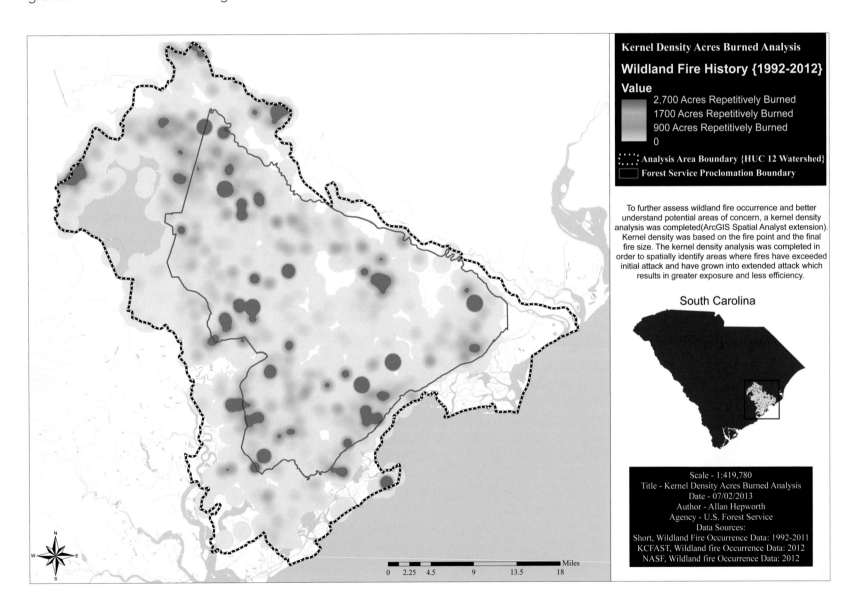

Kernel Density Acres Burned Analysis

Wildland Fire History {1992-2012}

Value

2,700 Acres Repetitively Burned
1700 Acres Repetitively Burned
900 Acres Repetitively Burned
0

Analysis Area Boundary {HUC 12 Watershed}
Forest Service Proclomation Boundary

To further assess wildland fire occurrence and better understand potential areas of concern, a kernel density analysis was completed(ArcGIS Spatial Analyst extension). Kernel density was based on the fire point and the final fire size. The kernel density analysis was completed in order to spatially identify areas where fires have exceeded initial attack and have grown into extended attack which results in greater exposure and less efficiency.

South Carolina

Scale - 1:419,780
Title - Kernel Density Acres Burned Analysis
Date - 07/02/2013
Author - Allan Hepworth
Agency - U.S. Forest Service
Data Sources:
Short, Wildland Fire Occurrence Data: 1992-2011
KCFAST, Wildland fire Occurrence Data: 2012
NASF, Wildland fire Occurrence Data: 2012

Miles
0   2.25   4.5   9   13.5   18

US Army Corps of Engineers (USACE)

Delano, Minnesota, Undocumented Levees Project

The map shows the discovery of undocumented flood control features to support flood damage reduction in Delano, Minnesota. A process was identified to support discovery of historic flood control feature by local governments. This resulted in increased efficiency and confidence in floodplain investigations to support flood damage reduction efforts.

Northern California Regional Intelligence Center (NCRIC)

NCRIC Market Street Security Camera

The Northern California Regional Intelligence Center (NCRIC) is a government program that helps safeguard the community by serving as a dynamic security nexus. To detect, prevent, investigate, and respond to criminal and terrorist activity, NCRIC disseminates intelligence and facilitate communications among federal, state, and local agencies and private sector critical infrastructure partners to help them take action on threats and public safety issues. The NCRIC created maps for the San Francisco Police Department showing the areas along Market Street that are covered by security cameras owned by private businesses. Market Street is a high traffic area for both transportation and commercial activity. It also functions as the primary parade route through the city. Maps such as these leverage an existing network of private security cameras to support special events and investigations along the Market Street corridor.

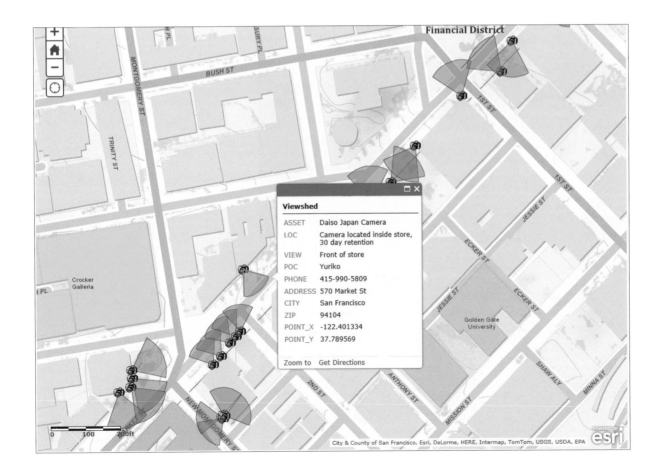

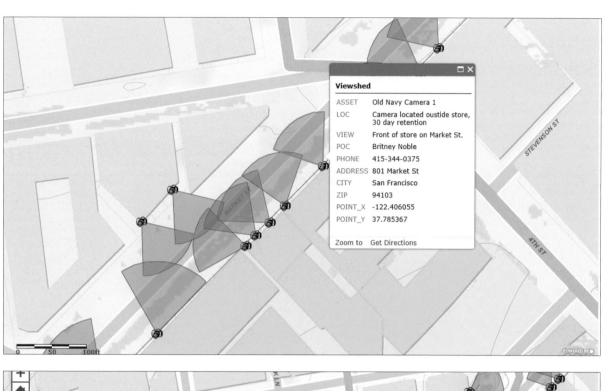

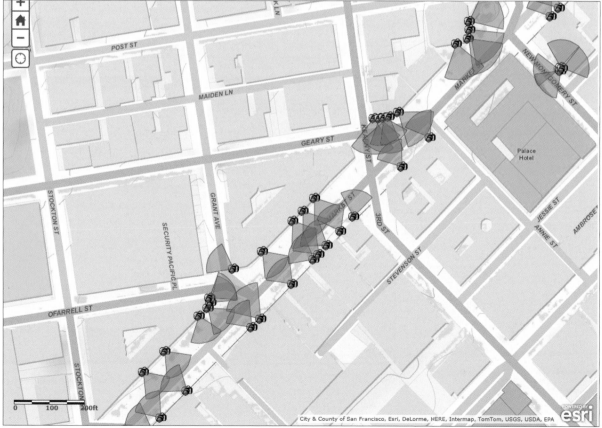

New York City Fire Department (FDNY) and PenBay Solutions

FDNY Super Bowl Event Management

With dozens of temporary venues and millions of visitors to the New York metro area, the weeks leading up to Super Bowl XLVIII on February 2, 2014, presented significant challenges to public safety planning. The Super Bowl brought fans from across the country to the New York metro area with hundreds of activities leading up to the game. The attractions provided entertainment to visitors and residents while posing significant challenges for the public safety organizations in charge of protecting these millions of people. To address the scale and complexity of the challenges—including the process of coordinating state, local, federal and private agencies—the New York City Fire Department worked with PenBay Solutions to roll out a new event management solution using InVision SecureSM software, a map-based public safety platform that leverages ArcGIS Online.

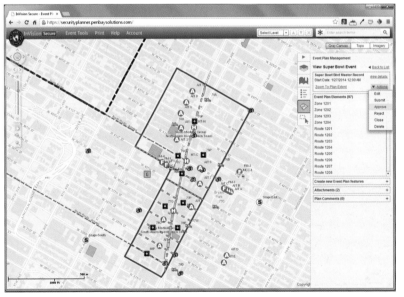

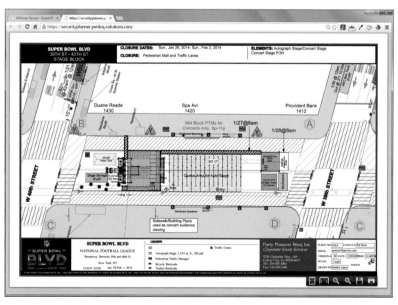

25

US Army Europe, Training Support Activity Europe and Tierra Plan

Army Training Support Activity Web App

The US Army Europe needed a way to Identify, acquire, manage and sustain training planers to provide state-of-the-art training support. TAP IN was developed as a GIS web application that provides a user-friendly means of locating and exploring training resources and facility scheduling data. The site includes information on all USAREUR training areas, ranges, simulators and devices. Each facility and location features links to training SOPs, handbooks and access maps. TAP IN also gives training units the ability to view current and future schedules of training facilities across Europe.

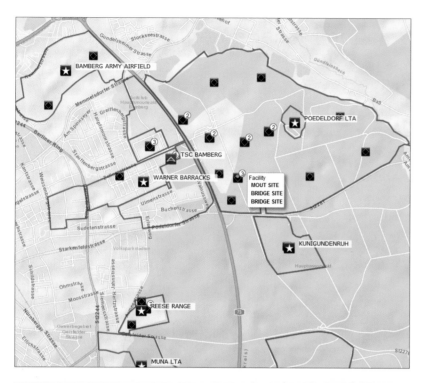

National Cancer Institute (NCI) and Information Management Services

Breast Cancer Mortality Rates

The National Cancer Institute studies cancer rates by geography and time. By using the same choropleth groups on both maps across time periods, one can see that breast cancer rates are falling. These maps help users find geographic outliers and patterns.

GIS Portal is a web-based station for interactive mapping and visualization of cancer related geospatial data. The portal combines GIS and science principles and tools to harmonize relatively large and multi-dimensional datasets, including population-based cancer statistics and behavioral, environmental, clinical, socioeconomic, and policy data at the county and state levels. The tools that comprise GIS Portal combine intelligent web maps with graphs, charts, tables, and text to inform, educate, and inspire users to generate research hypothesis.

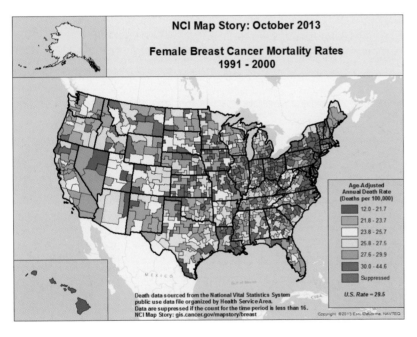

Virginia Department of Emergency Management

Virginia Department of Emergency Management Webpage

The Virginia Department of Emergency Management is charged with the coordination of the commonwealth's emergency preparedness, mitigation, response, and recovery efforts. GIS enables and facilitates collaboration through the use of geospatial data to protect Virginia and its localities from the impacts of natural and manmade hazards. This webpage is a collection of GIS maps used to enhance and communicate situational awareness regarding weather and natural hazards with potential to impact Virginia.

US Army Engineer School (USAES)

US Army Engineer School Maps,
Jalabad, Afghanistan

The US Army Engineer School is located at Fort Leonard Wood, Missouri, and provides training that develops a wide variety of engineering skills including combat engineer, bridging, construction, geospatial, topography, diving, and firefighting. These maps show the military aspect of the Jalabad, Afghanistan terrain developed at the school. The Jalabad Airport, operated by the US Armed Forces, is a key military installation in the region.

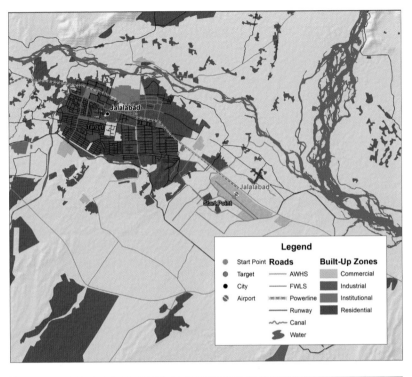

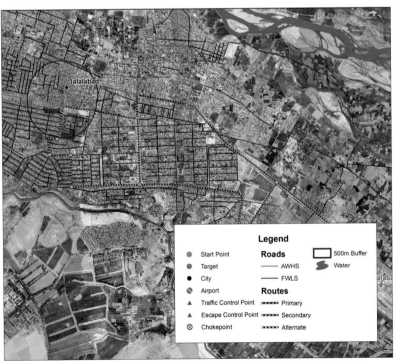

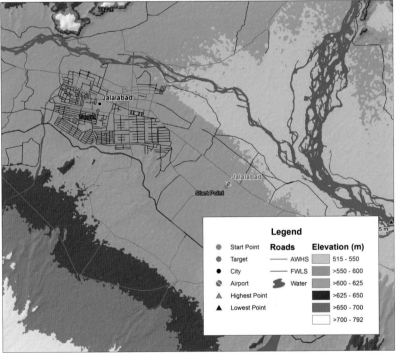

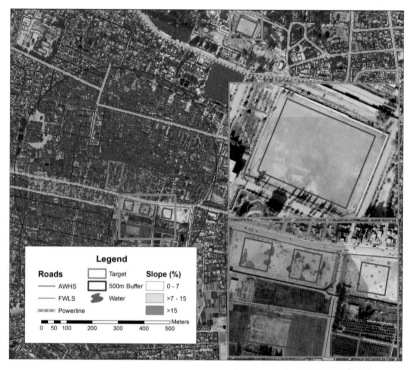

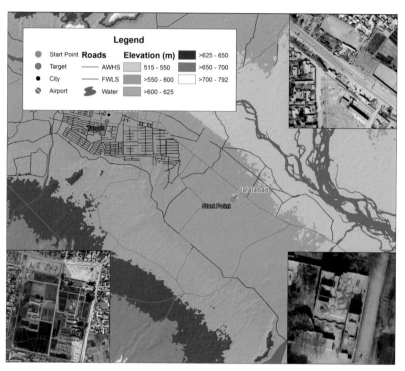

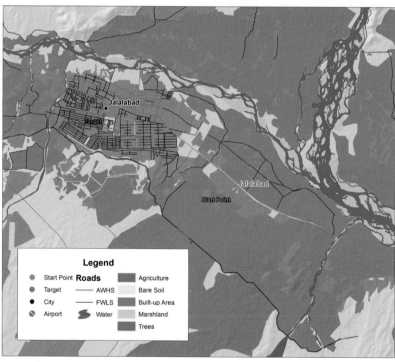

MAPPING
PUBLIC POLICY

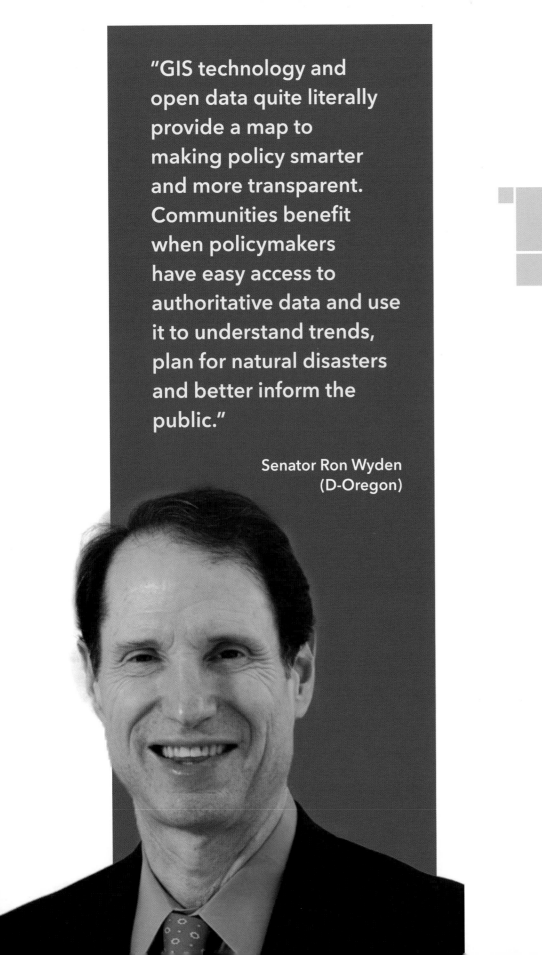

"GIS technology and open data quite literally provide a map to making policy smarter and more transparent. Communities benefit when policymakers have easy access to authoritative data and use it to understand trends, plan for natural disasters and better inform the public."

Senator Ron Wyden
(D-Oregon)

Mapping Public Policy

Policy makers at the highest levels of government use GIS to understand complex issues. Dynamic, informative maps give leaders an effective way to analyze problems, predict outcomes, and measure public sentiment by geography and point of view.

Maps also create a bridge between policy makers and the public. Online maps are a compelling new medium for reaching citizens because people intuitively understand data when it's presented in a map. Policy makers use maps to engage the public throughout the policymaking process—from a bill's infancy through implementation.

Whether the goal is connecting with the public, working is to gain momentum on a political initiative, or simply providing information on the world around us, elected officials at all levels of government use GIS to make smarter policy decisions and foster an informed citizenry.

US Senator James Risch's Map Gallery

United States Senate

Senator James Risch's Map Gallery is a great way for Idahoans to understand how their federal government is serving them. The Gallery provides information products that help the Senator's constituents understand their environment- both geographically and demographically. Through authoritative web map services, like the USGS active wildfire map, Idahoans can learn where active threats are. Through maps like the EPA cleanup site map, constituents can identify areas of pollution and learn where and how cleanup is taking place in their communities.

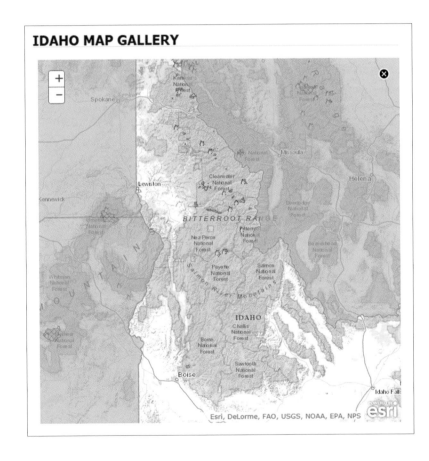

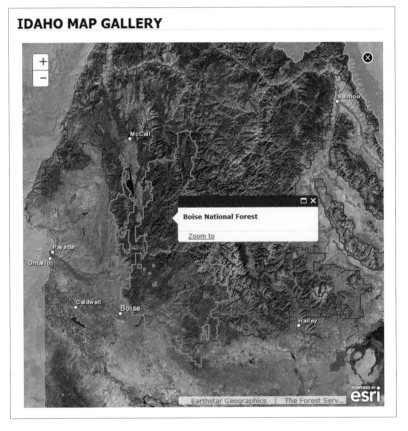

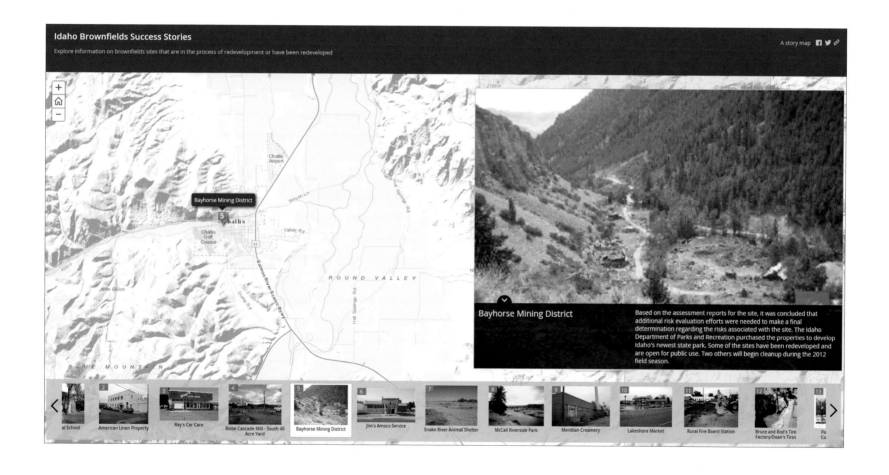

US Senator Ron Wyden's Chronic Disease Map

United States Senate

These county-level maps show the percentage of Medicare beneficiaries with chronic conditions. Sen. Ron Wyden (D-Ore.) used this map in a press release to demonstrate the need for his legislation to improve and standardize care for those with chronic illnesses. Constituents—and other legislators—can easily see how their county compares to the rest of the U.S. as well as how much variability there is across the country.

U.S. Senator Ron Wyden's Wildfire Map

United States Senate

This Map Journal displays wildfires currently burning in Oregon while explaining Sen. Ron Wyden's efforts to combat the problems of funding increasingly long fire seasons. The application uses multimedia to highlight the increasing severity and cost of wildfires and the Senator's plan to fund them like natural disasters instead of through federal agencies' regular budgets.

"GIS technology and open data quite literally provide a map to making policy smarter and more transparent. Communities benefit when policymakers have easy access to authoritative data and use it to understand trends, plan for natural disasters and better inform the public."—Sen. Ron Wyden (D-Oregon)

"Maps not only help make policy, but they can tell a story as well. Explaining and examining the impact and evolution of policy through maps is just as insightful as using GIS data to make legislative decisions."—Rebecca Steele (Digital Director, U.S. Senator Ron Wyden)

Access to Public Transit in California's 41st District

United States House of Representatives

As part of a larger report on transportation in his district, Congressman Mark Takano included a map of populations that are underserved by public transit. The analysis used 2010 Census Bureau data to show that about 23,000 households (almost 79,000 people) are more than a mile's walking distance from bus stops or rail stations in California's 41st district.

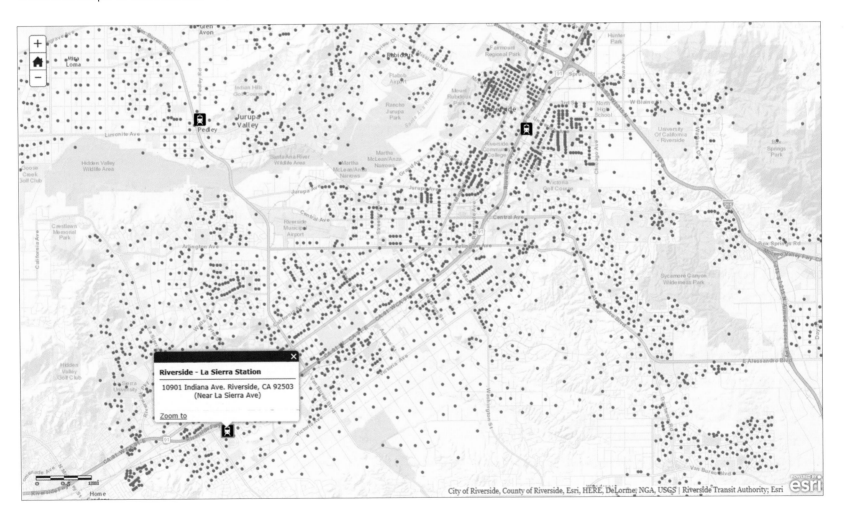

National Monuments Located Under Restricted Airspace

U.S. Library of Congress—Congressional Research Service (CRS)

This map is part of a series showing U.S. national monuments located under military airspace (i.e., prohibited, restricted, or Military Operations Area). The Congressional Research Service (CRS) identified a list of 107 national monuments recognized by the National Park Service, the Bureau of Land Management, and the U.S. Forest Service. Using data-driven software to create map, CRS analysts visually overlaid the national monuments (GIS layers) on images of the Federal Aviation Administration (FAA) Visual Flight Rules (VFR) charts. As a service unit within the Library of Congress, the Congressional Research Service (CRS) and its experts work with Congress through every step of the legislative process. Subject coverage is comprehensive and includes the use of GIS to analyze and illustrate issues of interest to our nation's lawmakers.

CRS services are available to every committee and Member of Congress and congressional staff. All CRS maps for Congress were reproduced with the authorization of the congressional requester.

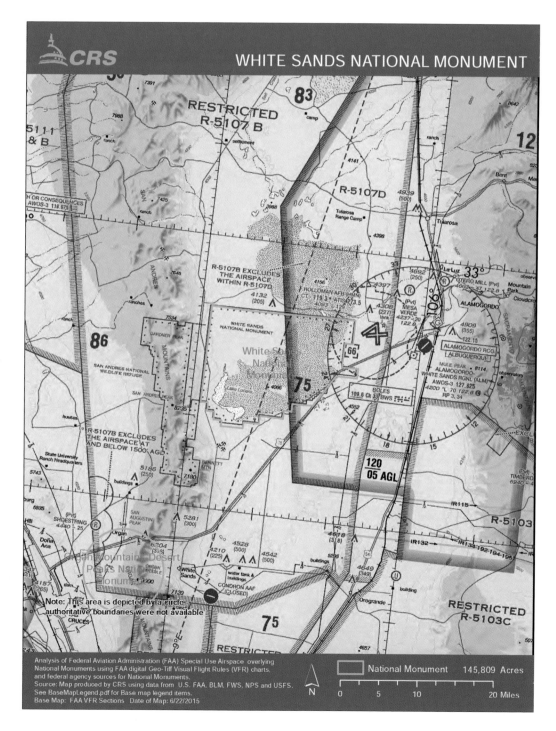

Disposition of U.S. Naval Petroleum Reserves

U.S. Library of Congress—Congressional Research Service (CRS)

This map shows the location and status of former Naval Petroleum Reserve (NPR) sites. A congressional client requested a brief history and status of the sites, beginning with the Naval Petroleum Reserves Production Act of 1976 (P.L. 94-258). The 1976 Act listed seven reserves and authorized their full commercial development. In 1977, the reserves were primarily transferred to and managed by the newly established Department of Energy (DOE). Using data from the U.S. Department of Energy, U.S. Department of Interior, the U.S. Geological Service, and Esri NatGeo World Street map, the Congressional Research Service (CRS) produced a visual history of the Reserves.

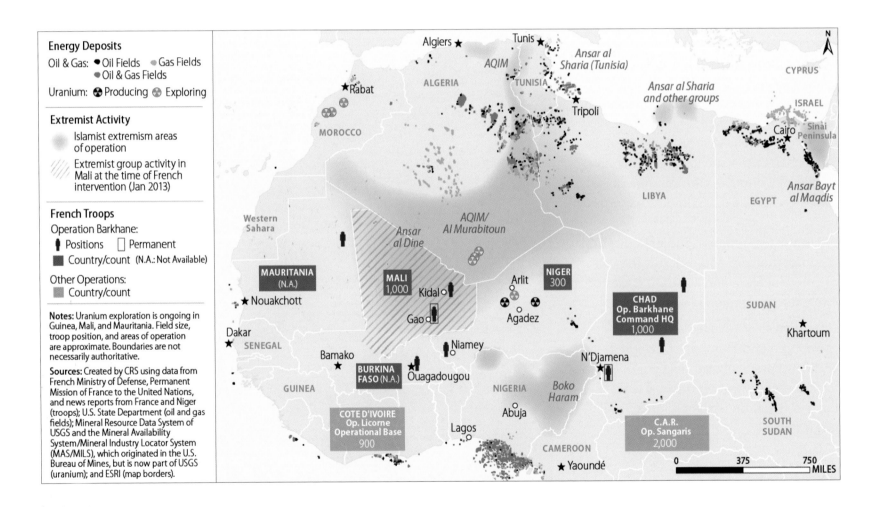

Map of the Sahel Region: Energy Resource Deposits, Extremist Group Areas of Operation, and French Troop Deployments

U.S. Library of Congress—Congressional Research Service (CRS)

On January 11, 2013, France launched Operation Serval in Mali, carrying out aerial bombing campaigns and ground operations targeting Islamist extremists.

Congressional Research Service (CRS) was asked for a map showing the energy resources in the region (including oil and gas fields and uranium mines), the areas of extremist operation, and the positions of French troops.

CRS analysts used data from the French Ministry of Defense, the Permanent Mission of France to the United Nations, and news reports from France and Niger to identify troop locations. U.S. State Department data was used to identify oil and gas fields.

The Mineral Resource Data System of USGS and the Mineral Availability System/ Mineral Industry Locator System (MAS/MILS)—now part of the USGS—identified uranium sites. Esri and Department of State Boundaries defined the map borders.

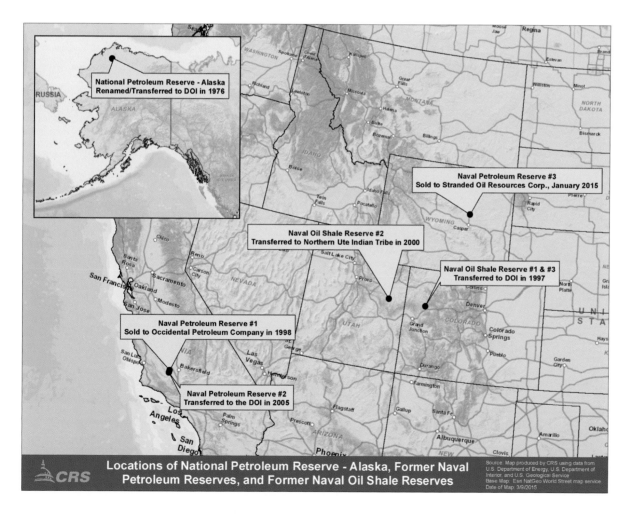

FOSTERING A HEALTHY NATION

"We must set our sights on a whole health care perspective that integrates geospatial data and insights to allow us to truly put consumers in control of their own health."

Susannah Fox, chief technology officer, US Department of Health and Human Services

Fostering a Healthy Nation

Nations around the world are prioritizing innovations to support improving the health and well-being of their people. GIS technology furthers those goals by providing insights about access to care, patterns of disease, and the availability of local resources to support mitigation strategies.

When health organizations consider serious issues like food safety, environmental hazards, and outbreaks, a geospatial approach aids planning, response, and trend analysis. With GIS, decision-makers can make the most of limited resources, analyze and visualize the current state of illness, predict the impact of interventions, and empower people to take action in support of their own health fostering smart communities that build smart nations.

US Food and Drug Administration (FDA)

Application of GIS in Work Planning Process: Service Area Analysis

GIS technology was used to conduct service area analysis (drive time and distance) to measure Food and Drug Administration (FDA) operation coverage of biologics firms in the Baltimore District. This project improves analysis of work planning and field operations including all geospatial elements that could depict field activities more accurately. Service area analysis (drive time) improves inspection coverage by efficiently reallocating resources and establishing priorities among firms and commodities at FDA.

Application of Geographic Information Systems (GIS)
Work Planning Process: Service Area Analysis

Alexis Novoa

Office of Regulatory Affairs; Division of Planning, Evaluation & Management; Program Evaluation Branch

Introduction

The U.S. Food and Drug Administration's (FDA) Office of Regulatory Affairs (ORA) is the lead office for all agency field activities. ORA inspects regulated products and manufacturers, conducts sample analyses of regulated products, and reviews imported products offered for entry into the United States. In pursuit of its mission, ORA also works with its state, local, tribal, territorial and foreign counterparts.

Geographic Information Systems (GIS) is a key resource for planning, evaluation, and management in FDA/ORA operations. GIS makes work planning more efficient and accurate, while maximizing the use of staff time and available resources. This project will use GIS technology to conduct Service Area analysis to measure FDA operation coverage of Biologics Firms in the Baltimore District.

Currently, ORA determines operation coverage based on multi-layer buffer zone polygons, which calculate linear distances without considering roads, terrain, and travel time. This method has challenges in reflecting actual conditions in the field. ORA is exploring new methods to improve analysis of field operations including all geospatial elements that could depict field activities more accurately.

Methodology and Tools

Utilized three different methods in ArcEditor to conduct spatial analysis:
1. Multi-layer Buffer Zone Polygons: This method generates a coverage area for an FDA facility by creating a buffer zone polygon based on a specified distance radius (1 mile, 5 miles, etc.).
2. Service Area Analysis (Distance): This method generates a coverage area for an FDA facility by creating a service area polygon which takes into account road length and type to calculate distance.
3. Drive Time Analysis: This method also creates service area polygons while considering road characteristics, but adds a time element (e.g. miles per hour) to generate a coverage area.

The number of firms within a coverage area and the elements (distance, road characteristics, and time) considered in each method were evaluated to determine which method is most useful.

Tool Used	Purpose
Excel Pivot tables	To create tables of firm counts
Proximity Tool - Multiple Ring Buffer	To represent the current analysis method
Esri ArcEditor - Network Analyst Tool - Service Area	To create Drive Time Polygons and Service Area Analysis
Select by Location	To identify the number of firms that fall in a particular distance parameter
Joint tables and export	To create a feature with the selected information

Findings

Service Areas	Total Firms	Percent
Buffer Rings	243	82%
Service Areas (Distance)	238	80%
Drive Time	260	87%
Baltimore District	298	100%

Percent of Firms Within Service Area Method

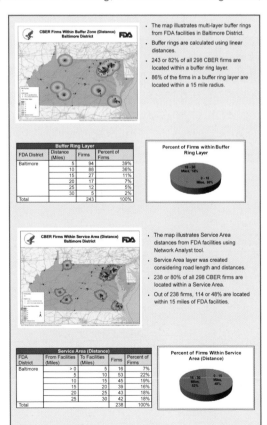

CBER Firms Within Buffer Zone (Distance) Baltimore District — FDA

- The map illustrates multi-layer buffer rings from FDA facilities in Baltimore District.
- Buffer rings are calculated using linear distances.
- 243 or 82% of all 298 CBER firms are located within a buffer ring layer.
- 86% of the firms in a buffer ring layer are located within a 15 mile radius.

Buffer Ring Layer

FDA District	Distance (Miles)	Firms	Percent of Firms
Baltimore	5	94	39%
	10	88	36%
	15	27	11%
	20	17	7%
	25	12	5%
	30	5	2%
Total		243	100%

Percent of Firms within Buffer Ring Layer

CBER Firms Within Service Area (Distance) Baltimore District — FDA

- The map illustrates Service Area distances from FDA facilities using Network Analyst tool.
- Service Area layer was created considering road length and distances.
- 238 or 80% of all 298 CBER firms are located within a Service Area.
- Out of 238 firms, 114 or 48% are located within 15 miles of FDA facilities.

Service Area (Distance)

FDA District	From Facilities (Miles)	To Facilities (Miles)	Firms	Percent of Firms
Baltimore	> 0	5	16	7%
	5	10	53	22%
	10	15	45	19%
	15	20	39	16%
	20	25	43	18%
	25	30	42	18%
Total			238	100%

Percent of Firms Within Service Area (Distance)

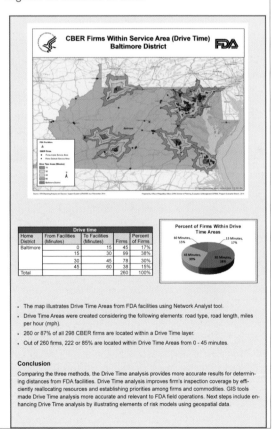

CBER Firms Within Service Area (Drive Time) Baltimore District — FDA

Drive time

Home District	From Facilities (Minutes)	To Facilities (Minutes)	Firms	Percent of Firms
Baltimore	0	15	45	17%
	15	30	99	38%
	30	45	78	30%
	45	60	38	15%
Total			260	100%

Percent of Firms Within Drive Time Areas

- The map illustrates Drive Time Areas from FDA facilities using Network Analyst tool.
- Drive Time Areas were created considering the following elements: road type, road length, miles per hour (mph).
- 260 or 87% of all 298 CBER firms are located within a Drive Time layer.
- Out of 260 firms, 222 or 85% are located within Drive Time Areas from 0 - 45 minutes.

Conclusion

Comparing the three methods, the Drive Time analysis provides more accurate results for determining distances from FDA facilities. Drive Time analysis improves firm's inspection coverage by efficiently reallocating resources and establishing priorities among firms and commodities. GIS tools made Drive Time analysis more accurate and relevant to FDA field operations. Next steps include enhancing Drive Time analysis by illustrating elements of risk models using geospatial data.

Evaluating the Wastewater Treatment Plant Impact on Shellfish

Growing Areas in Alabama's Mobile Bay

The Food and Drug Administration (FDA) provides guidance to state shellfish control authorities to establish prohibitive closure zones near wastewater treatment plant (WWTP) discharges. This minimizes molluscan shellfish exposure to health hazards posed by bacterial and viral pathogens present in wastewater effluents. From 2007 through 2009, the FDA conducted field investigations to assess the impacts of wastewater effluent from a large municipal WWTP that discharges into Alabama's Mobile Bay. GIS was used to geovisualize the zones of dilution, help determine the appropriate size of the prohibited zone, conditional zone, restricted zone and approved zone, and detect outlier data.

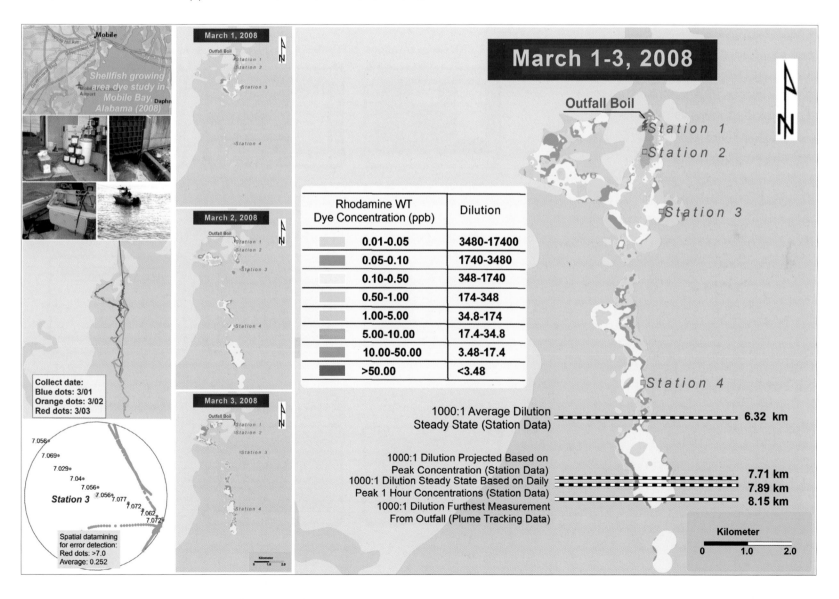

Rhodamine WT Dye Concentration (ppb)	Dilution
0.01-0.05	3480-17400
0.05-0.10	1740-3480
0.10-0.50	348-1740
0.50-1.00	174-348
1.00-5.00	34.8-174
5.00-10.00	17.4-34.8
10.00-50.00	3.48-17.4
>50.00	<3.48

FIT-MAP Mobile App

FIT-MAP automates the manual inspection process with mobile GIS functionality to increase the productivity of field investigators, improve data accuracy, and ultimately improve food safety. A paper-based inspection process resulted in poor data quality. FIT-MAP improved the quality of data to inform the agency's decision making and reduce the time needed to complete an inspection. Better data resulted in better decisions and faster data resulted in faster decisions.

FIT-MAP Mobile App

In 2013, the Center for Food Safety and Applied Nutrition (CFSAN) and Office of Regulatory Affairs (ORA) embarked on developing a field inspection pilot application called the Field Investigator's Tool with Mapping (FIT-MAP). FIT-MAP automates the manual inspection process and combines mobile data collection and Geographic Information System (GIS) functionality to increase the productivity of field investigators, improve data accuracy and ultimately improve food safety. FIT-MAP encompasses the foundation and framework for data collection and geocoding inspection features, through a secure and automated lightweight tablet.

The Mobile Device

- ITIM approved
- Smaller/lighter
- Ability to be sanitized
- PIV enabled
- Barcode reader
- GPS
- Touch screen/digitizer
- Camera
- Wi-Fi and 4G
- Windows 8.1
- Docking station

Toughpad FZ-G1 meets these requirements.

Inside the FIT-MAP App

Inside the FIT-MAP application, the FDA Investigator is able to select her pre-assigned farm inspection and see the location in an overview map (which is zoom enabled).

FIT-MAP Defines Locations

The FDA Investigator is able to create geo-referenced features (locations) as points, lines and polygons with underlining attribute information stored in a database.

FIT-MAP Asks Questions

The FIT-MAP questionnaire has a logic-based rule system. New questions appear based on the answers given to previous questions. It also allows for the investigator to associate each question and answer with a specific geo-referenced location.

FIT-MAP Records Samples

Samples are collected if directed by assignment, "For Cause", or by the investigator's own initiative (i.e. suspected rodent droppings).

FIT-MAP Tags Pictures

The FDA Investigator can create and attach pictures to samples and map locations, which ultimately get added to reports.

FIT-MAP Tags Files

The FDA Investigator can create and attach files to samples and map locations, which ultimately get added to reports.

FIT-MAP Tags Notes

The FDA Investigator can create and attach notes to samples and map locations, which ultimately get added to reports.

FIT-MAP Creates Forms

The FDA Investigator creates forms which are digitally signed and printed. All forms can be added to the final EIR reports.

FIT-MAP Inspection History

FIT-MAP gives investigators the ability to view inspection history reports from prior inspections.

FIT-MAP Field Tests
December 4, 2014

Initially, a cadre will perform 30 Investigations across the U.S. using FIT-MAP live in the field. The cadre will have access to the following:

FIT-MAP Training Curriculum
- Classroom Training
- Online Training Modules
- Video Clips

FIT-MAP Technical Support
- Helpdesk Phone Number
- Online Portal
- Email address

FDA GeoWeb

The Food and Drug Administration (FDA) needed an out-of-the-box solution to provide ready access to web maps and spatial data. Several groups in the agency were creating and purchasing spatial information, yet very few people even knew about FDA's GIS capability. With a small, yet hard-working team, the FDA was able to quickly establish a customized portal that allowed the agency to utilize spatial information with minimal training. The FDA GeoWeb quickly became the hub of spatial data in the agency.

Since its release in February 2013, employees from all of FDA's Centers and Offices have become registered users and many programs are reaping its benefits. OCM and the Office of Emergency Operations use FDA GeoWeb to monitor active emergencies that are fast moving and constantly changing. GeoWeb was used to provide situational awareness prior to and during the 2013 Presidential Inauguration and to determine impact to FDA regulated firms following the Moore, Oklahoma, tornado in May 2013.

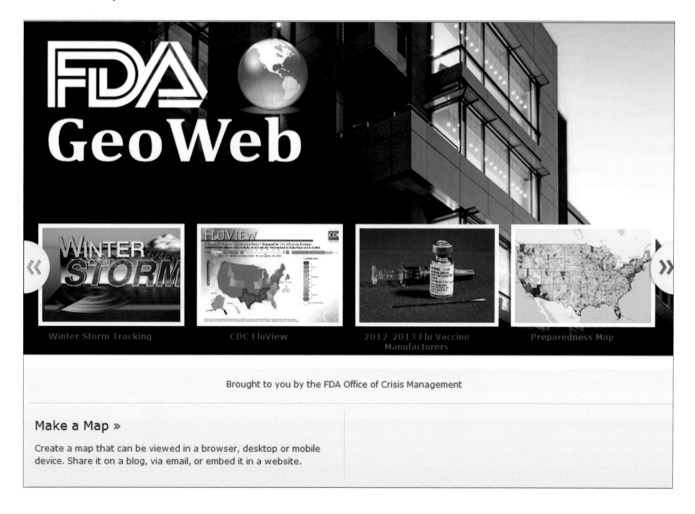

US Environmental Protection Agency (EPA)

Naturally Occurring Anthrax and the Geochemistry of American Soils

This map depicts areas that concurrently meet the hypothesized concentrations for elements that promote the persistence of anthrax (Bacillus anthracis). Certain geochemical elements in soil may promote anthrax persistence. The concentration of these elements in soil might vary in counties reporting animal anthrax cases compared to those that do not. The map depicts areas throughout the United States where concentrations of calcium (Ca), manganese (Mn), phosphorus (P) and strontium (Sr), were present in high concentrations. Elemental thresholds could be used to identify where a naturally occurring outbreak in animals might be more likely to occur than in other locations, with all other variables held constant.

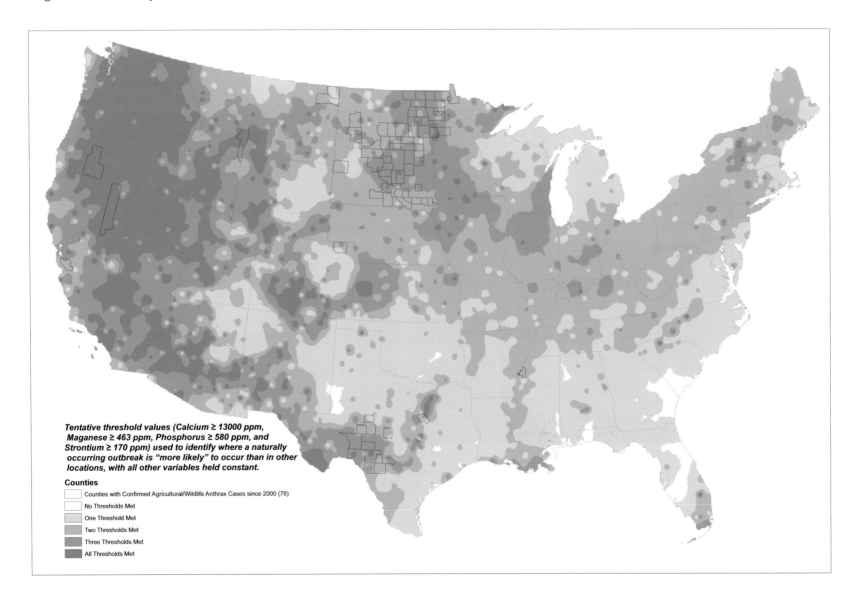

Tentative threshold values (Calcium ≥ 13000 ppm, Maganese ≥ 463 ppm, Phosphorus ≥ 580 ppm, and Strontium ≥ 170 ppm) used to identify where a naturally occurring outbreak is "more likely" to occur than in other locations, with all other variables held constant.

Counties

- Counties with Confirmed Agricultural/Wildlife Anthrax Cases since 2000 (78)
- No Thresholds Met
- One Threshold Met
- Two Thresholds Met
- Three Thresholds Met
- All Thresholds Met

US National Library of Medicine

Changes for the US National Library of Medicine TOXMAP

This poster describes the updating of TOXMAP, a 10-year-old ArcIMS-based GIS to ArcGIS Server and Flex. TOXMAP helps users visually explore data primarily from the US Environmental Protection Agency's Toxics Release Inventory (TRI) and Superfund programs. TOXMAP helps users create nationwide, regional, or local area maps showing where TRI chemicals are released on-site into the air, water, and ground. It also identifies the releasing facilities, color-codes release amounts for a single year or year range, and provides multi-year aggregate chemical release data and trends over time, starting with 1988. Maps can also show locations of Superfund sites on the National Priority List (NPL), listing all chemical contaminants present at these sites.

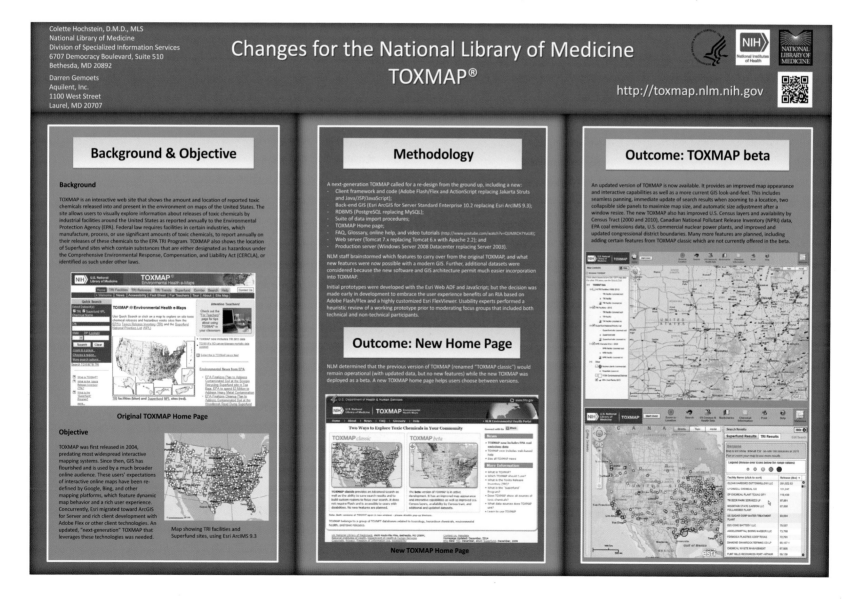

US Department of State, Humanitarian Information Unit

North Korea: Chronic Food Shortage Coping Mechanisms, Hillside Farming

Geospatial analysis using land cover and elevation data to determine how much agriculture in North Korea is on sloped lands. Sloped agriculture is an indicator that the people in that area are not obtaining enough food from the public distribution system. This map provides a better understanding of the food insecurity environment in North Korea and assistance for humanitarian partners to target appropriate food aid. The map analyzes food insecurity in a country with a very vulnerable population and significant risk of experiencing a complex humanitarian emergency.

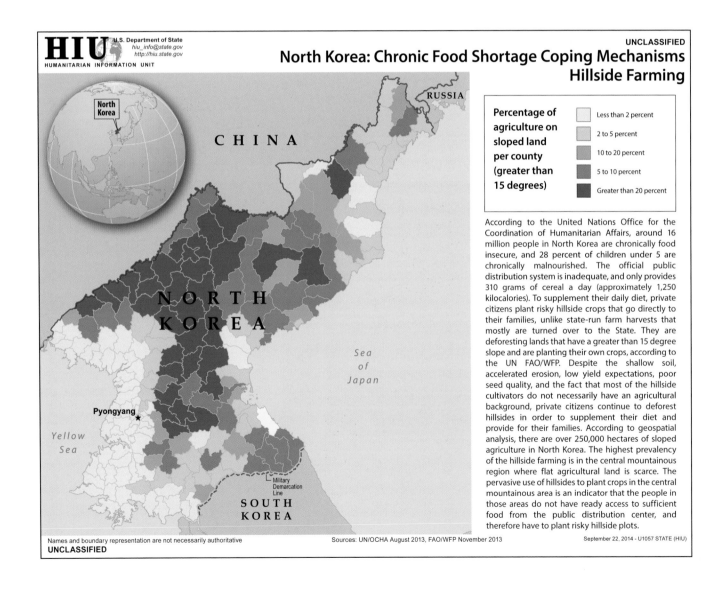

US Geological Survey (USGS)

Distribution of Lead (Pb): Soil C Horizon

The map shows the distribution of lead (Pb) in the soil C horizon (rock layer below the surface) in the conterminous United State interpolated from 4,857 sample sites collected and processed by the US Geological Survey (USGS). Lead is an element of concern for human and environmental health. This map defines areas with greater and lesser concentrations of Pb in soils of the conterminous United States better than previous ones. Determining and understanding the range of baseline Pb concentration and distribution levels is critical to recognize and understand changes in natural systems and human impacts on soil chemistry.

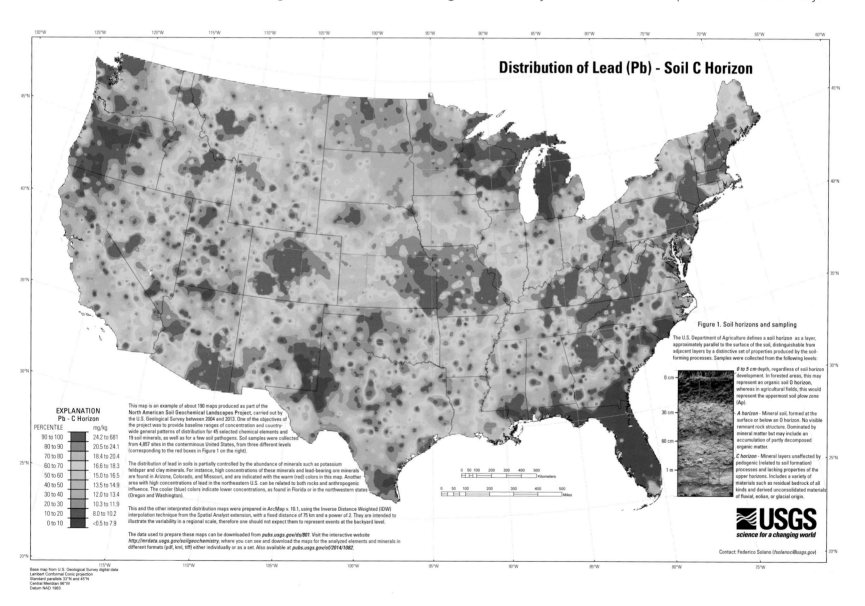

National Cancer Institute (NCI), Centers for Disease Control and Prevention (CDC), and Information Management Services

HPV and Cervical Cancer, Health Information National Trends Survey

The Health Information National Trends Survey (HINTS) data provides population estimates for variables that represent knowledge about certain cancer risk factors, screening tests and resources. This map addresses whether people think there is a link between human papillomavirus (HPV) and cervical cancer to help determine where program planners need to overcome cancer knowledge barriers for health communication. This map can be found on the statecancerprofiles.cancer.gov web site.

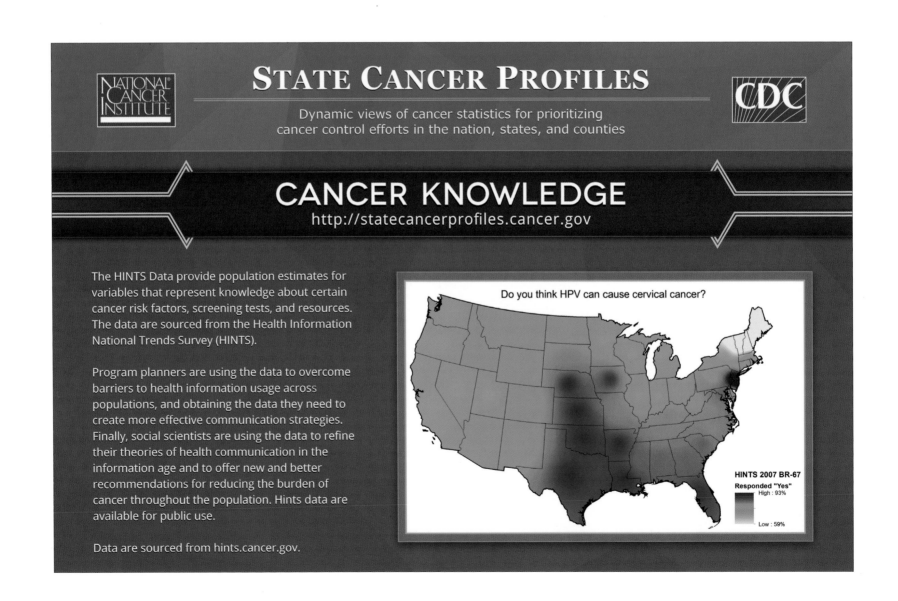

Centers for Disease Control and Prevention (CDC)

CDC Ebola Response Maps

(ID 6339) These are two maps show the cumulative and recent incidence of Ebola virus disease cases for the time period ending November 30, 2014.

(ID 6340/ 6342) These images are from a map book of the City of Nzerekore, Guinea, showing the map index and the two maps of grid cell 1 with and without imagery. The map book was used by CDC field epidemiologists for their investigations during the 2014 Ebola response. Each grid cell has two maps bounded next to each other. One map has the map with the satellite image and the other map is presented with no image.

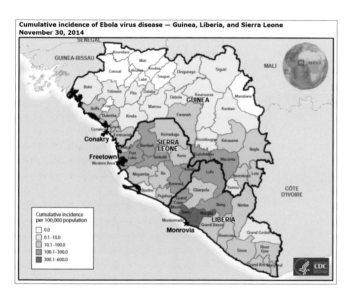

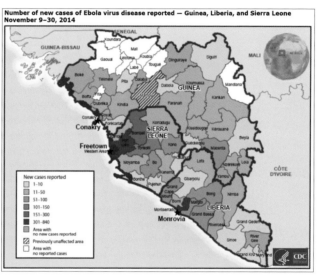

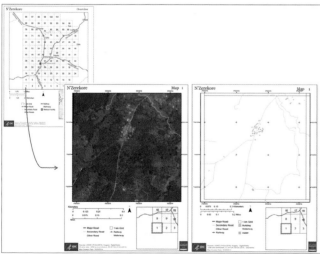

Heart Disease Death Rate

These maps show the heart disease death rate for the United States per 100,000 people from 2008 to 2010 and the death rate and poverty percentage in Ohio. The CDC also publishes heart disease mortality maps based on age, ethnicity, and other factors. The CDC developed these maps to give US residents and public health officials at the state, county, and local levels the tools they need to investigate where high rates of heart disease and stroke exist and who is mainly at risk. This is particularly important because the CDC report found that as many as 200,000 deaths that were attributed to cardiovascular disease could have been prevented.

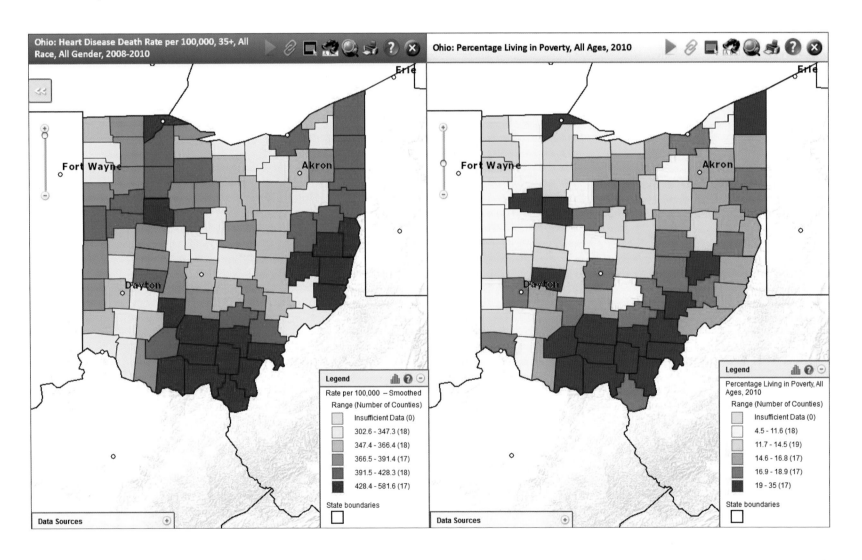

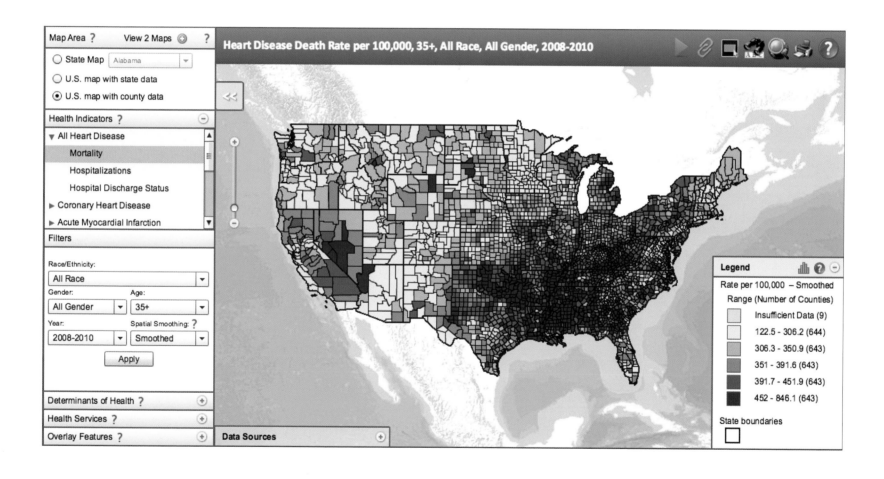

US Department of Health and Human Services, Office of Head Start

Children Living in Poverty

The Office of Head Start (OHS) is an agency within the Administration of Children and Families (ACF) under the US Department of Health and Human Services that promotes the school readiness of young children from low-income families through local Head Start and Early Head Start programs. These programs support the mental, social, and emotional development of children from birth to age 5. Service areas can span all fifty states, the District of Columbia, and US territories.

The map shown here was created to gain a better understanding of the poverty rates of counties across the United States for children under the age of 6. This map is part of a map journal featuring built-in queries to quickly identify counties in need across the nation. This map was created in ArcGIS online.

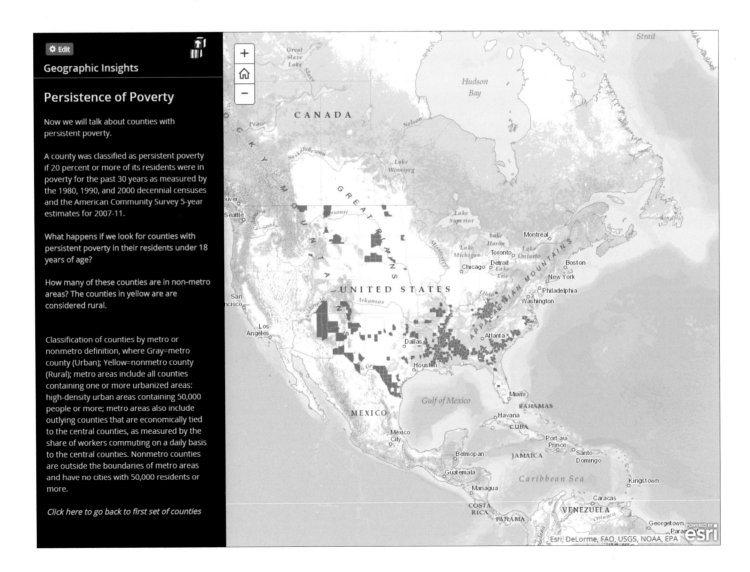

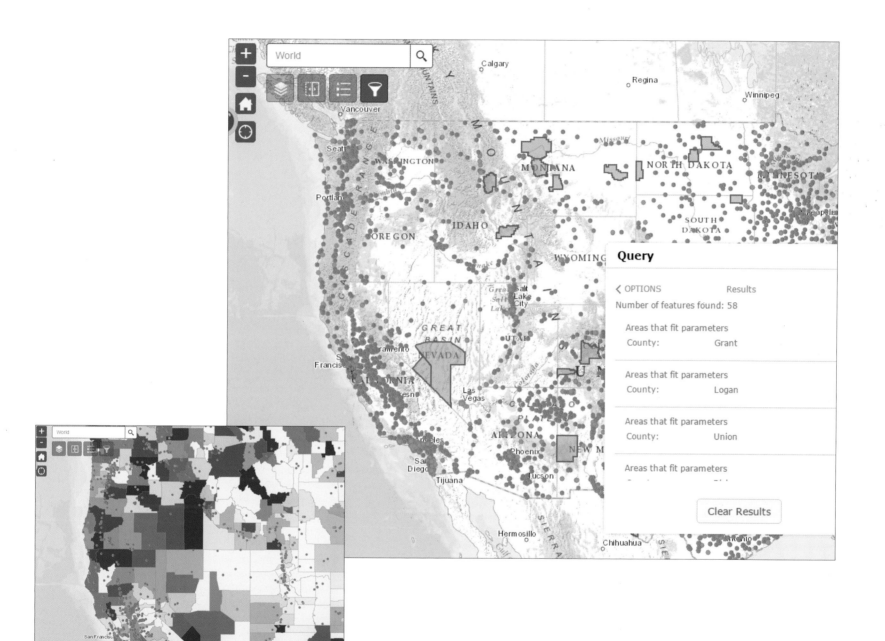

SUPPORT A RESILIENT ENVIRONMENT

"Forest Service leadership has long believed that the foundation of public policy is spatial in nature and its design is largely supported by GIS. Whether it's forest planning, restoration, or NEPA (National Environmental Policy Act), GIS is an important part of our day-to-day operations."

Zahid Chaudhry, GIS program manager, US Department of Agriculture Forest Service, Pacific Northwest Region

"Powerful analytical capabilities allow us to better examine characteristics such as wetland transition zones and provide a fundamental, quantifiable visualization tool to determine and plan for potential changes in salinity. As such, we can limit and monitor potential impacts to significant wetland resources and more importantly, evolve our awareness of coastal resilience concepts while planning for improvements to our nation's critical navigation infrastructure."

Molly Reif, research geographer, ERDC-EL and US Army Corp of Engineers

Supporting a Resilient Environment

Smart nations and communities are resilient nations and communities. They prepare for and recover quickly from disaster or hardship. Scientists, arborists, conservationists, and more use ArcGIS in research, analysis, and dissemination of critical information to support a resilient environment.

A location-based system of engagement is critical to informing and promoting wide-ranging and impactful policies that protect our natural resources, and in understanding the long-term environmental impacts of the decisions we make today. To enhance public outreach on these issues, governments have developed innovative GIS-based approaches to communicating the work being done to reduce the impacts of a changing environment.

From the depths of the ocean floor off the coast of Oregon to the urban forests of Washington, D.C., and the polar bear population in the Arctic, the ArcGIS platform delivers information that addresses local and global issues affecting our daily lives. Using geography to integrate information from collection activities, observations and explorations, and big data systems allows scientists, policy makers, and the public to make more complete and informed decisions for a more resilient environment.

National Oceanic and Atmospheric Administration (NOAA)

NOAA OAR (Ocean and Atmospheric Research) PMEL (Pacific Marine Environmental Laborator) Earth-Ocean Interactions Program

Navigating under the ocean at 1,500 meters of depth presents many challenges such as the lack of GPS signals and visible light. ArcGIS software enables scientists on research expeditions to more effectively see the ocean floor to conduct their scientific research and exploration. After using a remotely operated vehicle to gather samples, make observations, and deploy experiments, the integration of the information into a working GIS effectively helps manage operations at sea. This 2013 ROV Jason dive at Axial Seamount, 300 miles off the Oregon coast, explored in detail a 2011 lava eruption. The red bacterial mat formed in the aftermath of the eruption was highly sought after by a pharmacologist who was able to take a sample after its initial discovery on the dive. Scientists were also able to place complex instrumentation on top of a hydrothermal vent in an energy-generation experiment.

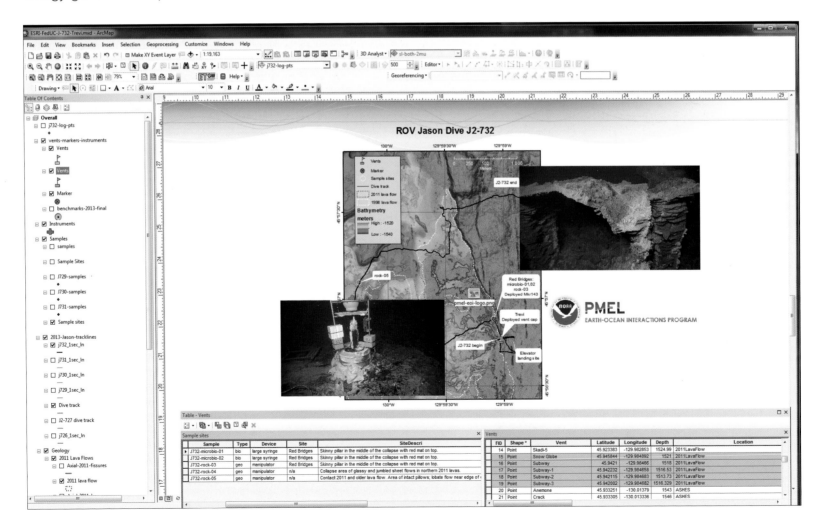

US Army Corps of Engineers (USACE)

Cooper River Critical Wetland Habitat

This is a map of critical wetland habitats classified from WorldView-2 imagery along the Cooper River, north of Charleston, South Carolina. The map shows the habitats as well as geo-corresponding habitat relative abundance (bar graph on the right) to further illustrate critical transition zones between saltwater, brackish, and freshwater marshes trending south to north along the river.

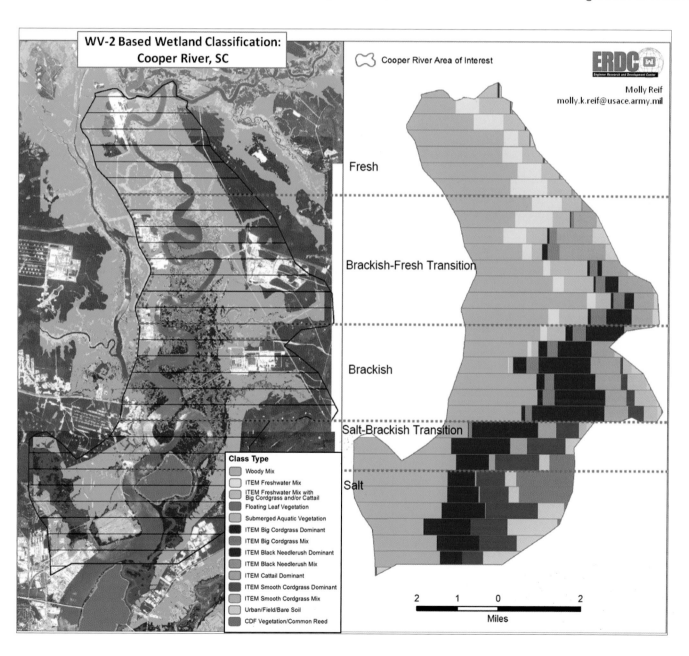

US Department of Agriculture (USDA) National Agriculture Imagery Program (NAIP)

Northern Minnesota NAIP Imagery

This dataset contains 3-band natural color imagery from the National Agricultural Imagery Program (NAIP). NAIP acquires digital orthoimagery during the agricultural growing seasons in the continental United States. A primary goal of the NAIP program is to enable availability of orthoimagery within one year of acquisition. NAIP imagery is available for distribution within sixty days of the end of a flying season and is intended to provide current information of agricultural conditions in support of US Department of Agriculture farm programs.

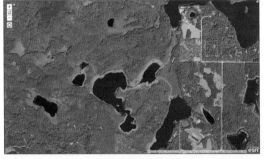

Bureau of Ocean Energy Management (BOEM) and Geodynamics

Potential Wind Energy Lease Blocks Offshore Cape Fear, North Carolina

North Carolina has some of the best conditions to support offshore wind energy in the southeastern United States. Geodynamics was contracted to assess and characterize the seafloor off Cape Fear in support of researchers at University of North Carolina at Chapel Hill's Institute for Marine Science and the National Oceanic and Atmospheric Administration's Center for Coastal Fisheries Habitat Research.

Geodynamics developed a detailed survey plan to conduct preliminary environmental site characterizations using side scan and multibeam sonar to detect objects on the seafloor. The side scan sonar imagery displays seafloor type characteristics for an area of potential wind energy lease-blocks. These surveys will help refine potential lease areas to reduce conflict among ocean uses and minimize environmental impacts off the North Carolina coast.

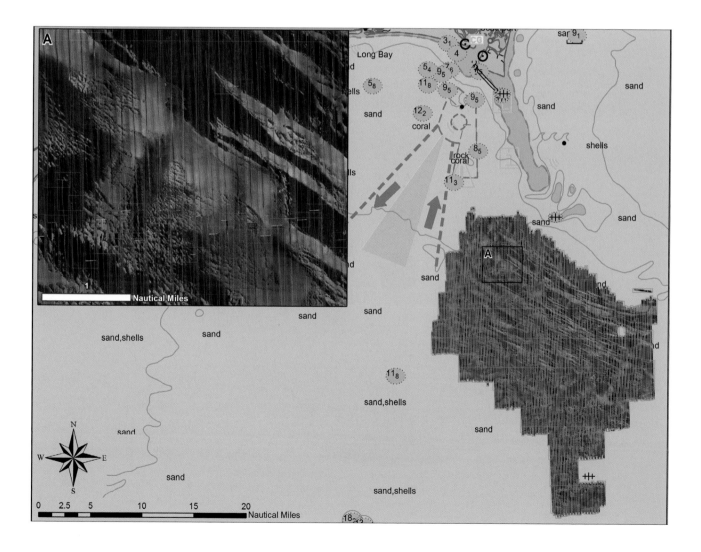

United Nations Environment Programme

Sea Ice Concentration and the State of the Polar Bear

The sea ice map tries to illustrate the effect of climate change on the Arctic ice and this helps in formulating policies that will help to mitigate the effects of climate change to the environment. The other map illustrates the effect of the Arctic ice on the population of the polar bears. Polar bears are not evenly distributed throughout the Arctic, nor do they comprise a single nomadic cosmopolitan population, but rather occur in nineteen relatively discrete subpopulations. There is, however, an uncertainty about the discreteness of the less studied subpopulations, particularly in the Russian Arctic and neighboring areas, due to very restricted data on live capture and tagging. The total number of polar bears worldwide is estimated to be 20,000 to 25,000.

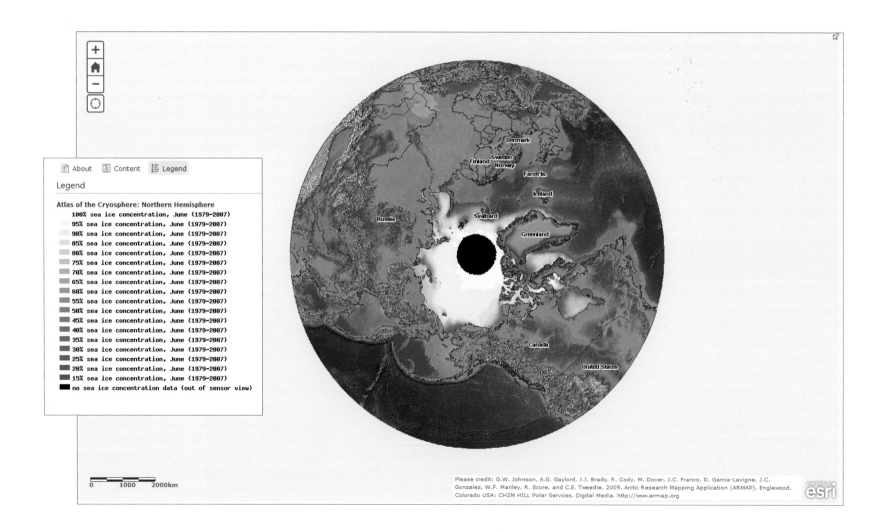

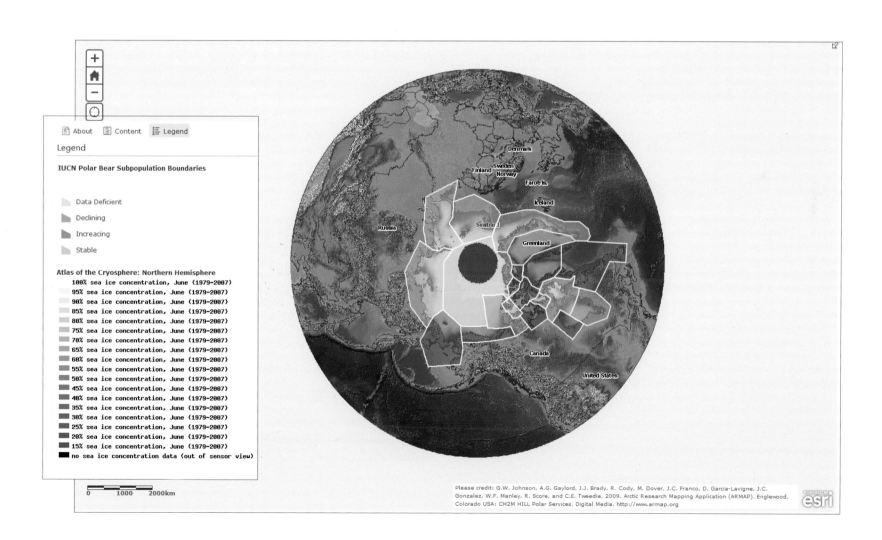

About　　Content　　Legend

Legend

IUCN Polar Bear Subpopulation Boundaries

Data Deficient

Declining

Increacing

Stable

Atlas of the Cryosphere: Northern Hemisphere

100% sea ice concentration, June (1979-2007)
95% sea ice concentration, June (1979-2007)
90% sea ice concentration, June (1979-2007)
85% sea ice concentration, June (1979-2007)
80% sea ice concentration, June (1979-2007)
75% sea ice concentration, June (1979-2007)
70% sea ice concentration, June (1979-2007)
65% sea ice concentration, June (1979-2007)
60% sea ice concentration, June (1979-2007)
55% sea ice concentration, June (1979-2007)
50% sea ice concentration, June (1979-2007)
45% sea ice concentration, June (1979-2007)
40% sea ice concentration, June (1979-2007)
35% sea ice concentration, June (1979-2007)
30% sea ice concentration, June (1979-2007)
25% sea ice concentration, June (1979-2007)
20% sea ice concentration, June (1979-2007)
15% sea ice concentration, June (1979-2007)
no sea ice concentration data (out of sensor view)

0　　1000　　2000km

Please credit: G.W. Johnson, A.G. Gaylord, J.J. Brady, R. Cody, M. Dover, J.C. Franco, D. Garcia-Lavigne, J.C. Gonzalez, W.F. Manley, R. Score, and C.E. Tweedie, 2009. Arctic Research Mapping Application (ARMAP). Englewood, Colorado USA: CH2M HILL Polar Services. Digital Media. http://www.armap.org

71

US Energy Information Administration (EIA)

EIA Energy Maps

The US Energy Information Administration (EIA) is part of the US Department of Energy and a principal agency of the US Federal Statistical System responsible for collecting, analyzing, and disseminating energy information to promote sound policymaking, efficient markets, and public understanding of energy and its interaction with the economy and the environment. EIA programs cover data on coal, petroleum, natural gas, electric, renewable and nuclear energy. Examples of EIA maps are shown here.

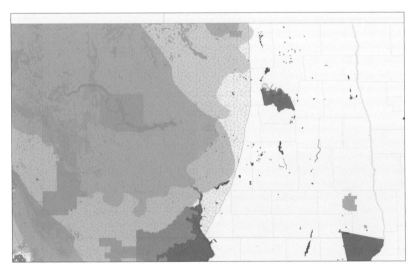

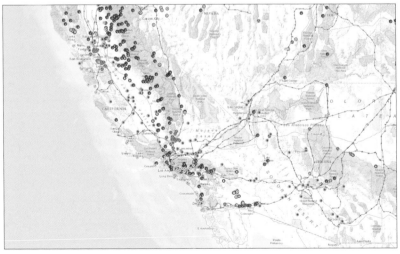

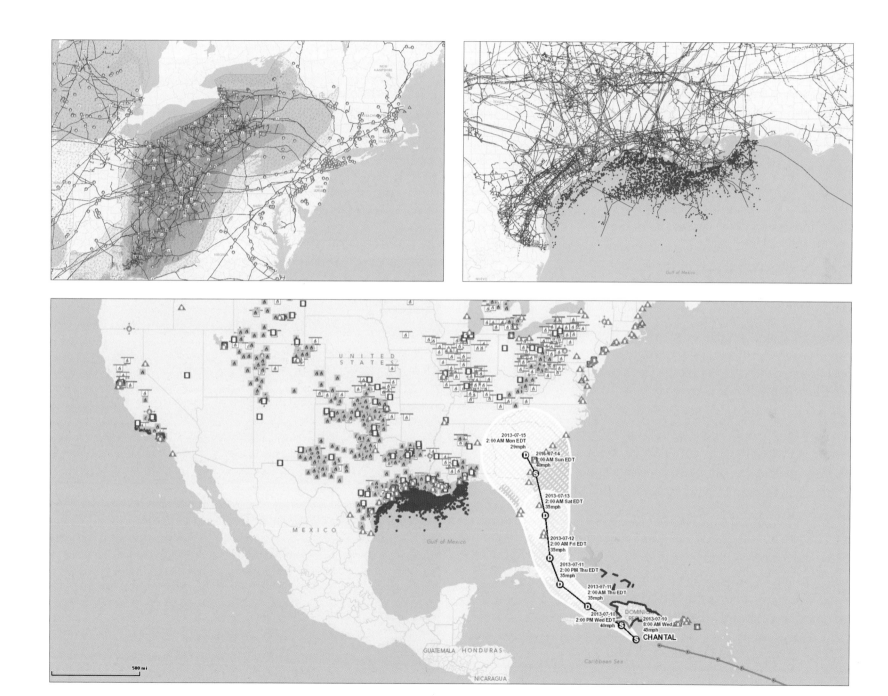

2013-07-15
2:00 AM Mon EDT
29mph

2013-07-14
2:00 AM Sun EDT
35mph

2013-07-13
2:00 AM Sat EDT
35mph

2013-07-12
2:00 AM Fri EDT
35mph

2013-07-11
2:00 PM Thu EDT
35mph

2013-07-11
2:00 AM Thu EDT
35mph

2013-07-10
2:00 PM Wed EDT
40mph

2013-07-10
8:00 AM Wed
45mph

CHANTAL

UNITED STATES

MEXICO

Gulf of Mexico

GUATEMALA HONDURAS

NICARAGUA

Caribbean Sea

DOMINICAN REP.

500 mi

District of Columbia Department of Transportation, Urban Forestry Administration

Urban Forest in Washington, DC

The Urban Forestry Administration (UFA) installed over 8,000 new street trees this past season and involved a half-dozen crews installing more than 100 new trees per day. The operations dashboard allowed UFA to monitor the progress of this effort in real time, across the entire city. By sharing the dashboard, everyone from the city council to the general public were able to monitor the tree planting deployment.

Story maps have helped UFA convey the importance of trees in the city, while also highlighting efforts to enhance this shared resource. They create a space where the public can relate to the trees that surround them and better understand this resource. Story maps also enable the UFA to communicate its efforts to preserve and enhance the urban forest of Washington, DC.

ArcGIS has enabled arborists to easily and accurately capture detailed information on facilities like bioswales, rain gardens, Flexi-Pave Sidewalks, and storm water retention ponds installed within the Department of Transportation's right of way. Creating this detailed inventory has assisted UFA in performing routine maintenance, as well as sharing the distribution of these sites with the public.

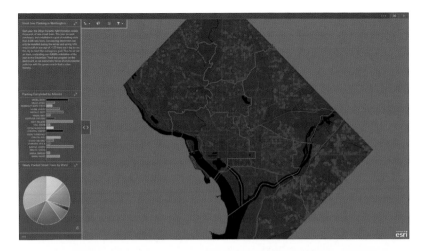

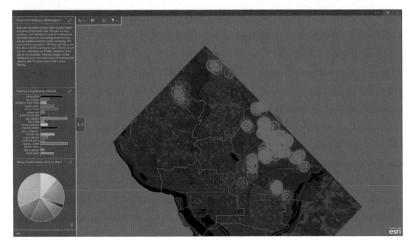

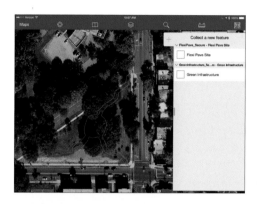
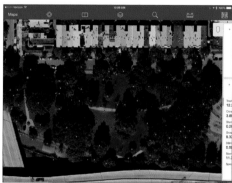

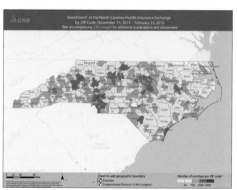
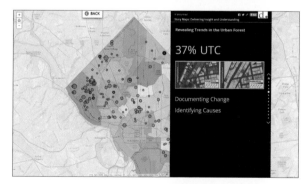

OPTIMIZING NATIONAL MAPPING AND STATISTICS

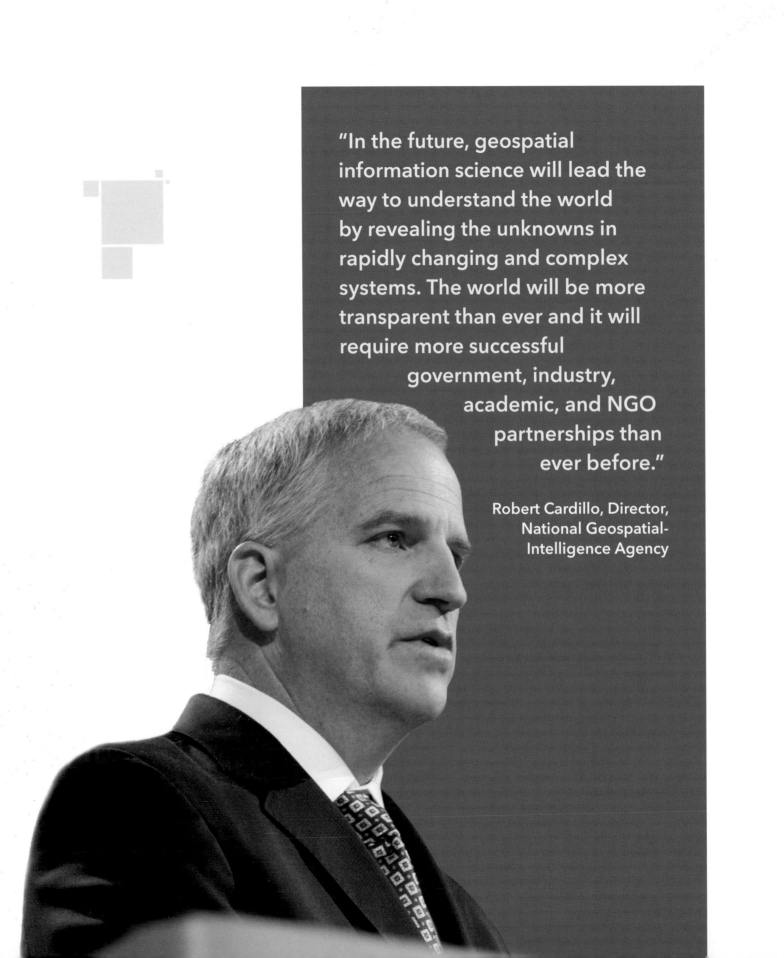

"In the future, geospatial information science will lead the way to understand the world by revealing the unknowns in rapidly changing and complex systems. The world will be more transparent than ever and it will require more successful government, industry, academic, and NGO partnerships than ever before."

Robert Cardillo, Director, National Geospatial-Intelligence Agency

Optimizing National Mapping and Statistics

National Mapping and Statistical organizations deliver geographic content and information products that provide the foundation information for national governments to meet new challenges within their nation and communities.

Authoritative national map and land information support smarter and more effective management of public lands, conservation or development of natural resources and ensure the protection of lives and property. Maps and charts underpin smart transportation initiatives and safety of navigation. Statistical information, such as population census, business and agriculture data, is used to drive economic development and improve citizen access to services.

GIS provides the platform for a smart nation. It enables governments, industry, academia and citizens the ability to discover, analyze and share information. Smart nations utilizing GIS technology can make intelligent decisions from their nation's foundational content to successfully meet the challenges and priorities they face in the 21st Century, to improve lives and business.

US Bureau of Land Management (BLM)

BLM's National Conservation Lands: Protecting Large Landscapes

The US Bureau of Land Management is seeking to advance a landscape approach to management for its National Conservation Lands. The maps display the large landscapes protected by the National Conservation Lands and their location in context with urban areas and illustrate using landscape-scale data to ID restoration priorities. The maps provide a profound realization to the agency and public that the BLM is already engaged in landscape-scale conservation near urban centers. The maps help move the conversation about landscape-scale management in the agency from theory to practice.

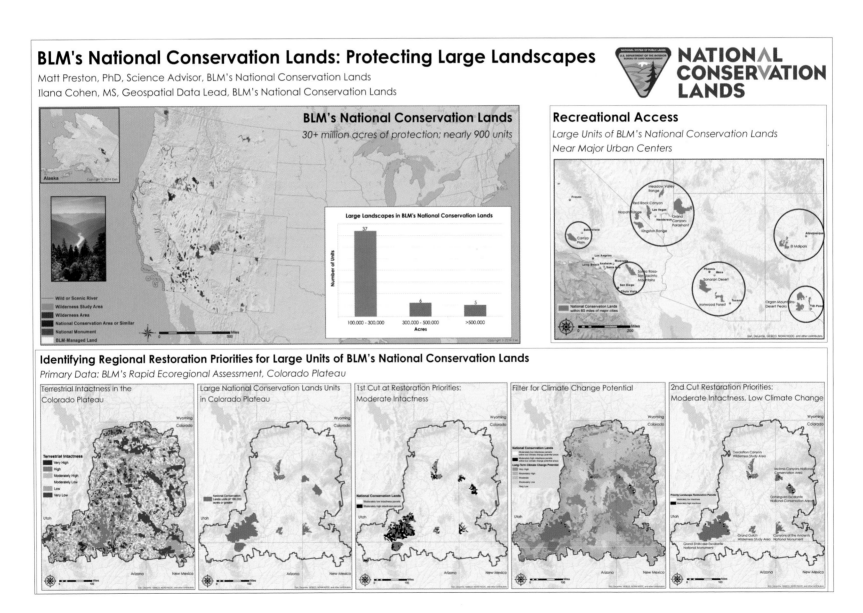

Public Information Map– Assayii Lake Fire

This public information map for Assayii Lake Fire in June 2014 shows locations of great importance to the Navajo Nation. During a public meeting, the map was presented in a way that the Navajo people could easily recognize where the locations important to them were in reference to the fire perimeter. The maps also included the locations in the Navajo language. The result was better communication with the Navajo people for suppression activities for the Assayii Lake Fire.

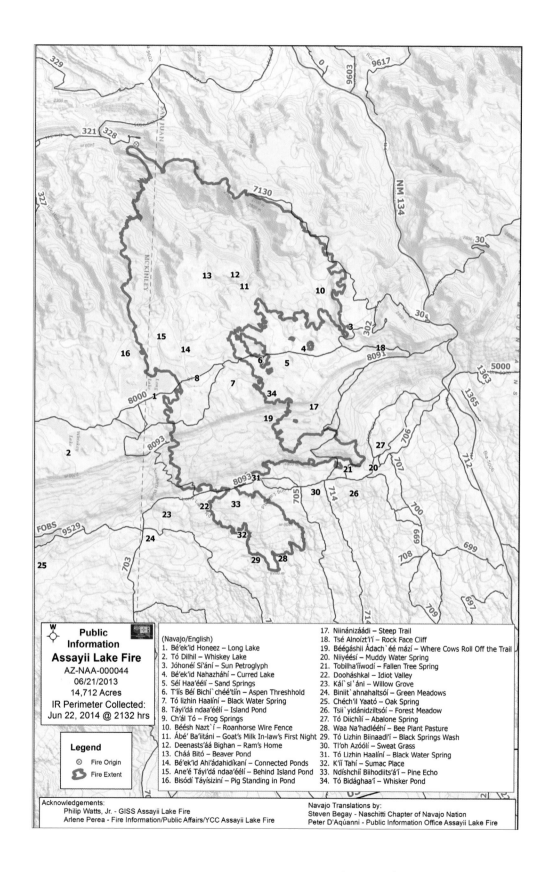

Public Information Assayii Lake Fire
AZ-NAA-000044
06/21/2013
14,712 Acres
IR Perimeter Collected:
Jun 22, 2014 @ 2132 hrs

Legend
⊗ Fire Origin
Fire Extent

(Navajo/English)
1. Bé'ek'id Honeez – Long Lake
2. Tó Dilhil – Whiskey Lake
3. Jóhonéí Sí'ání – Sun Petroglyph
4. Bé'ek'id Nahazháhí – Curred Lake
5. Séí Haa'éélí – Sand Springs
6. T'íís Béí Bichi`chéé'tíín – Aspen Threshhold
7. Tó lizhin Haalíní – Black Water Spring
8. Táyi'dá ndaa'éélí – Island Pond
9. Ch'ál Tó – Frog Springs
10. Béésh Nazt`í – Roanhorse Wire Fence
11. Ábé' Ba'iitáni – Goat's Milk In-law's First Night
12. Deenasts'áá Bighan – Ram's Home
13. Cháá Bitó – Beaver Pond
14. Bé'ek'id Ahi'ádahidíkaní – Connected Ponds
15. Ane'é Táyi'dá ndaa'éélí – Behind Island Pond
16. Bisódí Táyísizini – Pig Standing in Pond
17. Niinánizáádi – Steep Trail
18. Tsé Alnoízt'í'í – Rock Face Cliff
19. Béégáshii Ádach`éé mází – Where Cows Roll Off the Trail
20. Niiyéésí – Muddy Water Spring
21. Tobilha'ííwodí – Fallen Tree Spring
22. Dooháshkal – Idiot Valley
23. Káí`si`áni – Willow Grove
24. Biniit`ahnahaltsóí – Green Meadows
25. Chéch'il Yaató – Oak Spring
26. Tsii`yidánidziltsóí – Forest Meadow
27. Tó Diichílí – Abalone Spring
28. Waa Na'hadlééhí – Bee Plant Pasture
29. Tó Lizhin Biinaadl'í – Black Springs Wash
30. Tl'oh Azóólí – Sweat Grass
31. Tó Lizhin Haalíní – Black Water Spring
32. K'íí Tahí – Sumac Place
33. Ndíshchíí Biihodliits'á'í – Pine Echo
34. Tó Bidághaa'í – Whisker Pond

Acknowledgements:
Philip Watts, Jr. - GISS Assayii Lake Fire
Arlene Perea - Fire Information/Public Affairs/YCC Assayii Lake Fire

Navajo Translations by:
Steven Begay - Naschitti Chapter of Navajo Nation
Peter D'Aqúanni - Public Information Office Assayii Lake Fire

US Department of Agriculture (USDA) Natural Resources Conservation Service (NRCS)

Characterization of Playa Hydrology on the Southern High Plains

This map describes the use of high resolution elevation data to characterize the hydrology of the Southern High Plains, including the quantification and assessment of ecological functions of playa lakes which are round hallows in the ground present certain times of the year. Most playas fill with water only after spring rainstorms when freshwater collects in the round depressions of the otherwise flat landscape of West Texas, Oklahoma, New Mexico, Colorado, and Kansas. High-resolution digital elevation data accurately maps catchment areas without the need for intensive ground surveys. This map is useful for conservation planning and wildlife habitat analysis.

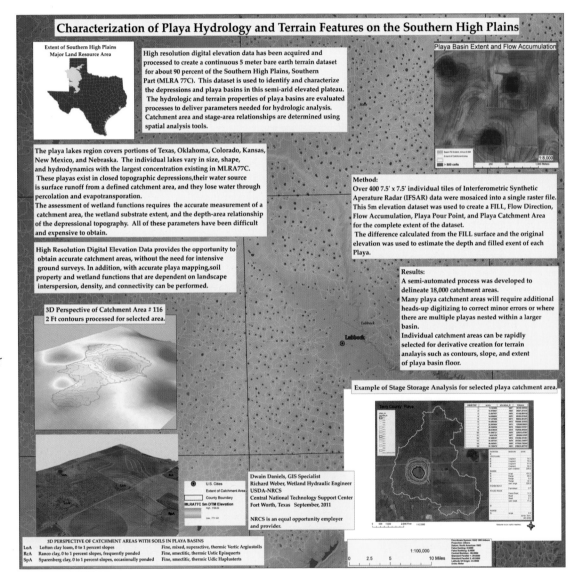

Appalachian Mountain Club

Land Conservation in Maine: 1994–2014

This map displays the change over time in land conservation in the state of Maine from 1994 to 2014. The 1994 map addresses the amount of land conserved while the 2014 map displays the amount of land that has been conserved in the last twenty years by federal, state and local governments, as well as conservation organizations in this region. This map is used to display how much work has been done in land conservation in the last twenty years and where work still needs to happen.

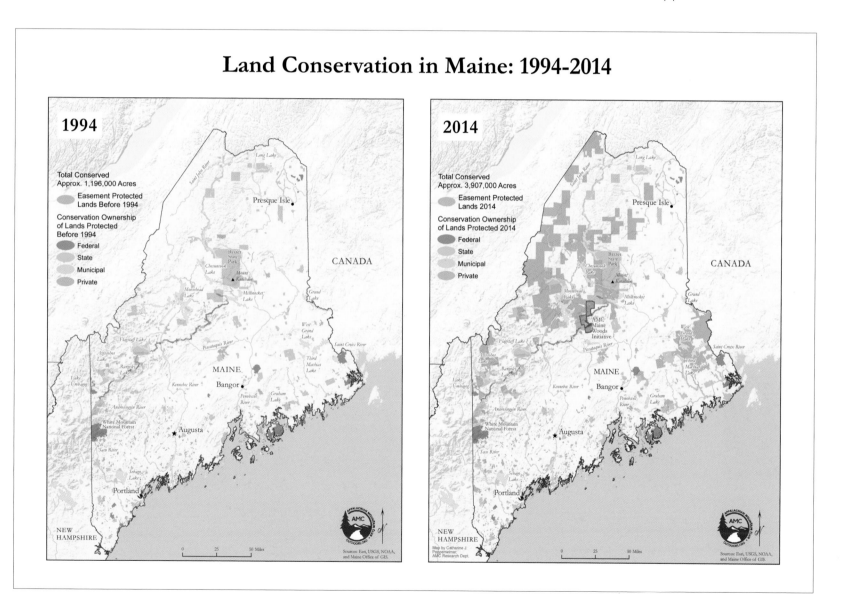

US Geological Survey (USGS)

Quaternary Geologic Map of the North-Central Part of the Salinas River Valley and Arroyo Seco

Arroyo Seco exposes a sequence of dated strath terraces (which result from either a stream or river downcutting through bedrock) that exhibit the effects of long-term climate change in a tectonically active environment. This map addresses the geologic and climatic influences that formed the strath terraces. Tectonic movements result in a slowly emerging coast line but the regional effects of climate change are interpreted as the driving mechanism for erosion and aggradation in Arroyo Seco. This analysis provides the US Geological Survey with a long-term paleoclimate record in a tectonically active regime.

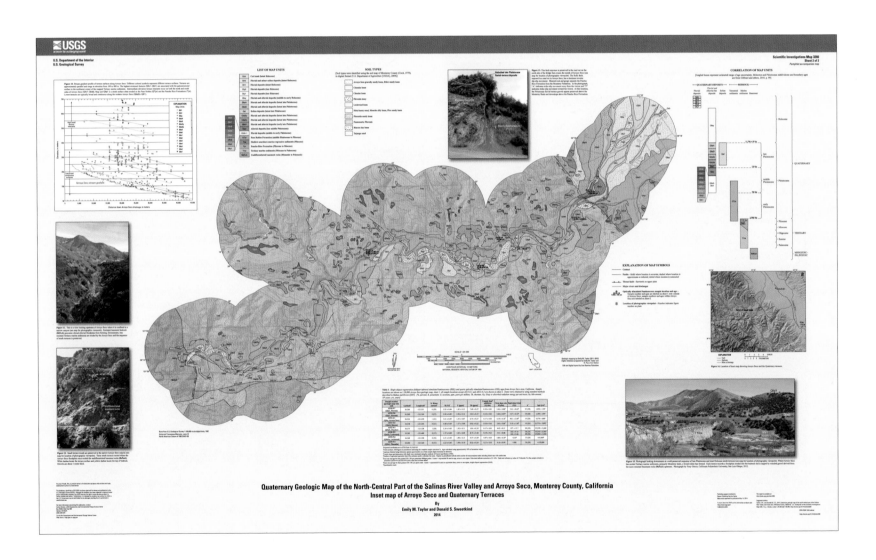

3D Elevation Program: Status of 3DEP Quality Data

The 3D Elevation Program uses the US Interagency Elevation Inventory to measure progress and improve lidar data acquisition coordination. Discovering existing elevation data can be difficult because many entities may have collected data. The US Interagency Elevation Inventory (USIEI) actively seeks to improve awareness of existing data. The US Geological Survey-led 3D Elevation Program needs to understand where elevation data already exist in the United States to be able to measure progress and to plan and coordinate acquisitions with partners. The USIEI multi-agency partnership saves the 3D Elevation Program time and money by increasing awareness and access to existing data and aiding in program planning.

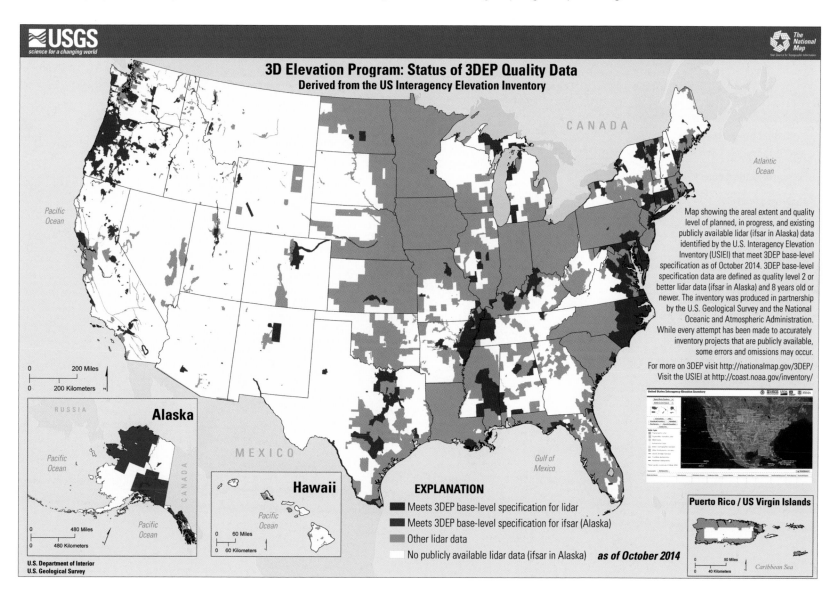

US Census Bureau

Census Data Mapper Homeowner Vacancy Rate

The Census Data Mapper is a web mapping application that provides users with a simple interface to view, save, and print county-based demographic maps of the United States and Puerto Rico. The statistical data is from the 2010 Census and the geographic boundaries are from the Census Bureau's Topologically Integrated Geographic Encoding and Referencing (TIGER) database. Statistical data that is used by governments and the private sector to make important decisions can be explored geographically and thematic maps can be created using the Census Bureau's standard cartographic design and layout. This application improves access to census data, and allows users to quickly and easily create customized thematic maps displaying data from the 2010 Census.

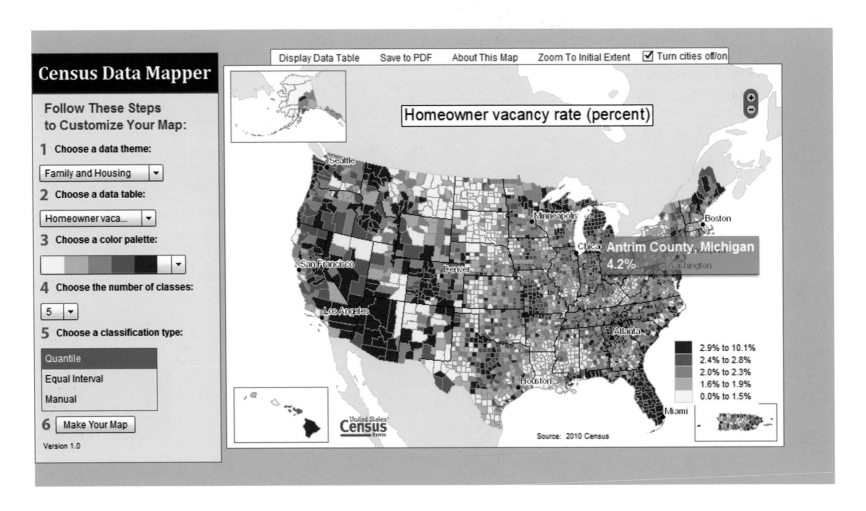

Census Bureau's TIGER Map Viewer and Web Mapping Services

TIGERweb is the Census Bureau's TIGER map viewer and web mapping services. Members of the public use these maps to view geography relationships or use the web mapping services with the Census data API to create their own census data application with the most current data. The TIGERweb is a web-based mapping application that allows users to view and query TIGER data. It contains legal and statistical geographic boundaries as well as roads, railroads, and hydrography features. In addition, the TIGERweb viewer includes attribute information, to include 2010 Census and Census 2000 population and housing unit counts.

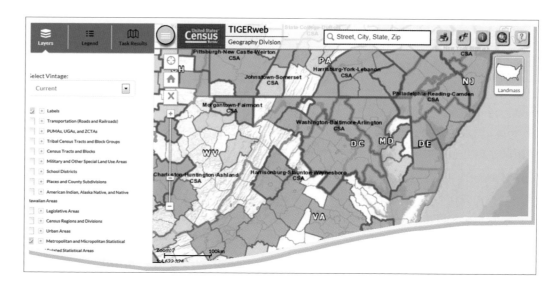

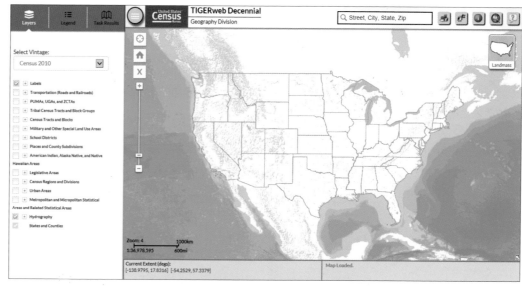

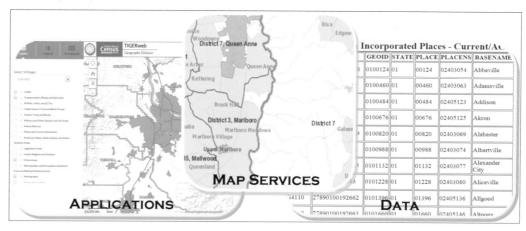

Cabo Verde National Statistics Institute

Census 2010 Project

For a census, it is necessary to determine statistical units or information collection areas (enumeration areas or census block). To this end, it is necessary to make a map that covers the entire territory, divided into statistical units (census blocks) where the inquirer collects information avoiding duplication or omission.

The census blocks should be easily recognizable, so its limits must be clearly visible and represented in a clearly visible scale. In this context, GIS is a fast and reliable tool for census cartography. Before using Esri technology, the census cartography was made on paper, at varying scales and sometimes hindered the recognition of enumeration areas. Also, its implementation took a long time, required a lot of workers, and the information could not be produced for small areas.

ArcGIS improved the representation of enumeration areas and allowed better control of work during the census operation. And, census information could be produced for smaller geographical levels and specific levels, all georeferenced.

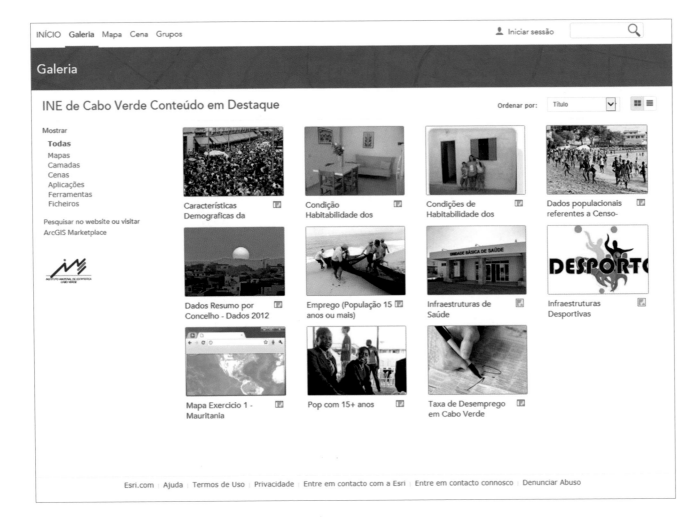

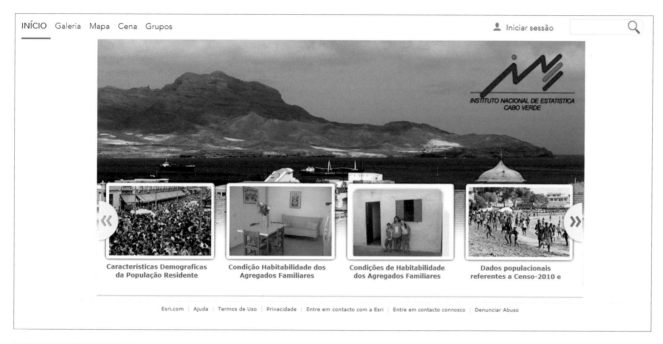

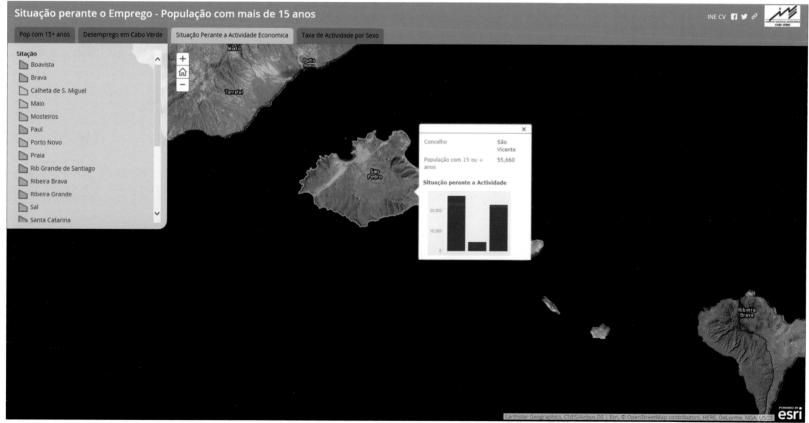

Dutch Kadaster

Time Travel App

Dutch Kadaster is the Netherland's cadastre, land registry, and mapping agency. In 2015, Kadaster celebrated 200 years of Dutch topographic mapping with its Time Travel App topotijdreis.nl. In this website one can easily travel through 200 years of maps for the country at all available scales. Places can be found by address and themes like reclamation, urbanization and land development are clearly recognizable in the maps. Kadaster also used a Story Map to tell this story which can be read at http://arcg.is/1V7b1Pb

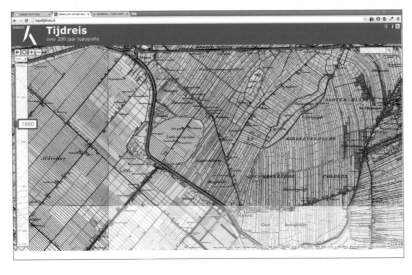
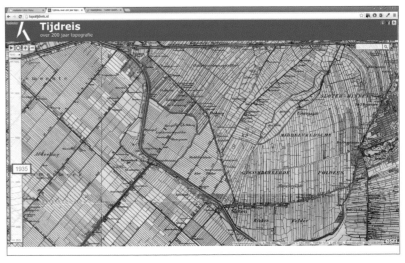
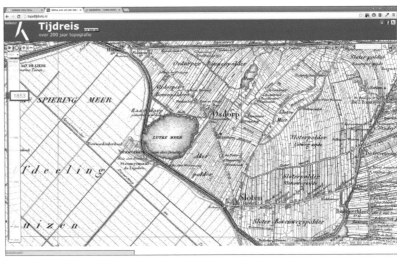
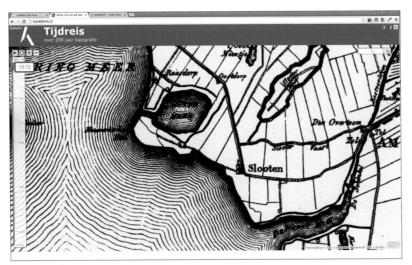

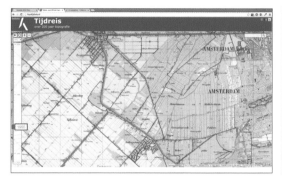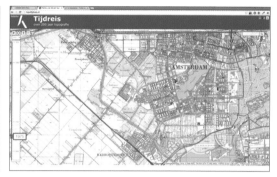

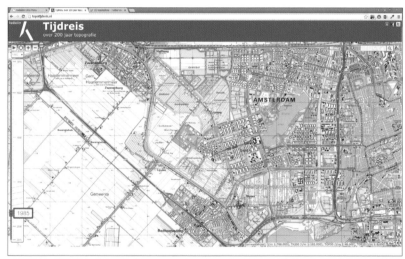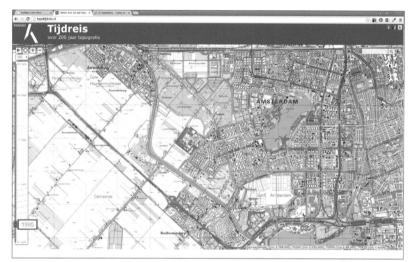

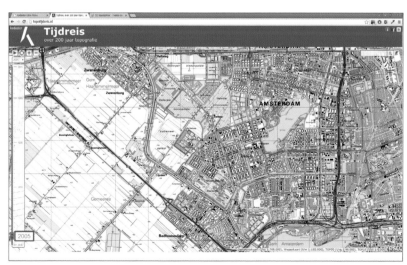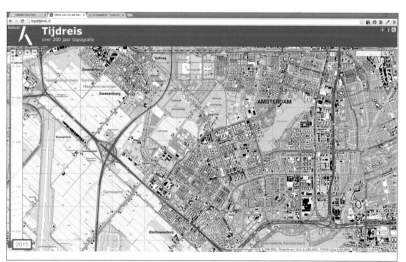

ELEVATING EDUCATION

Elevating Education

First and foremost, the business of education is teaching, learning, the creation of new knowledge through research, and public service. A geographic perspective supports the kinds of active, problem-based learning that engage students and teachers. GIS strengthens teaching and learning, career readiness, and interdisciplinary research.

Educational institutions also use GIS to maintain physical campuses. These "learning spaces" are a mix of real property, associated physical assets, and supporting infrastructure. From a functional perspective, many larger campuses and school districts resemble small cities. As such, they require stewardship—planning, management, maintenance, and sustainability. Given the persistent economic challenges that confront all levels education, the efficiencies that can be gained by running smarter campuses are powerful.

University of Maryland—Campus Webmap

http://maps.umd.edu/map/

The University of Maryland (UMD) wanted the ability to query building-to-building pedestrian directions for both shortest distance and wheelchair accessible routing. Working with input from the UMD's President's Commission on Disability Issues, Facilities Management staff was tasked to add current accessible pathways to the Campus Webmap along with shortest distance routing. Facilities Management, with the assistance of student interns from the Department of Geographical Sciences, created a full campus pedestrian transportation network connecting all university buildings.

The inclusion of the campus building navigation options was met with an enthusiastic response from the campus community. Additionally, an optional elevation profile viewer has been integrated using Esri World Elevation services to explore elevation change over the selected route, which has also garnered a lot of positive attention. Facilities Management routinely updates the pedestrian transportation network when new construction occurs and regenerates the routing available in the campus map on a quarterly basis.

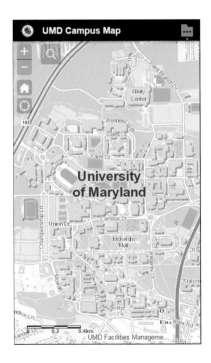
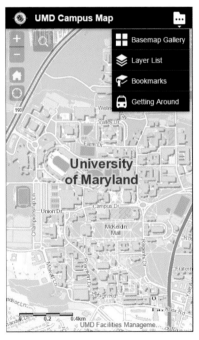

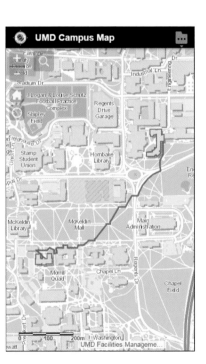

UMD's Interactive Campus Map

NOW FEATURING:

- **Enhanced support for mobile devices**
- **Search and view campus features including: buildings, parking lots, and other campus services and amenities**
- **Create, print, and share your own map view**
  - **Find routes to and from campus buildings**
  - **Access to ShuttleUM and WMATA bus stop arrival information and live tracking of bus locations**

MAPS.UMD.EDU

National Center for Education Statistics and Blue Raster

MapED: School Bullying

This map unites Census and US Department of Education National Center for Education Statistics (NCES) datasets with GIS mapping capabilities to allow the general public to visualize the data with ease. By displaying the data on a national map down to the district or school level, it makes it accessible to a much wider audience in an approachable way. A publicly accessible mapping platform that allows users to narrow down hundreds of census demographic indicators over multiple years, starting with a topic of interest, and then visualize it on a map.

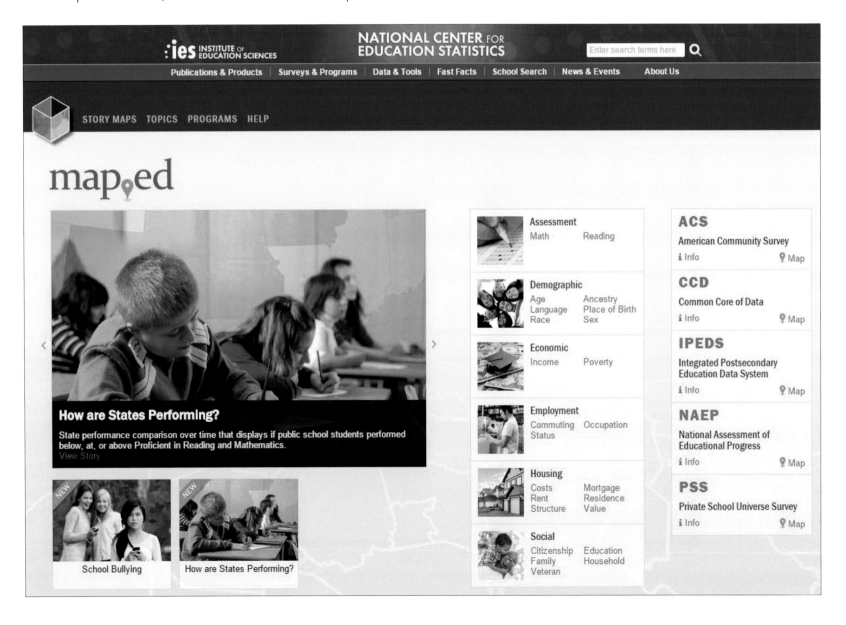

MapED: Public School Student Performance on the National Assessment of Educational Progress

The Nation's Report Card interactive map illustrates the achievement-level results of the 2013 National Assessment of Educational Progress (NAEP), but not the results of any prior years. The story map displays a historical timeline comparison of public school student performance on the NAEP from 2003 to2013. This provides policymakers, education professionals, and the public a way to better understand patterns in how states are performing on national assessments in reading and mathematics.

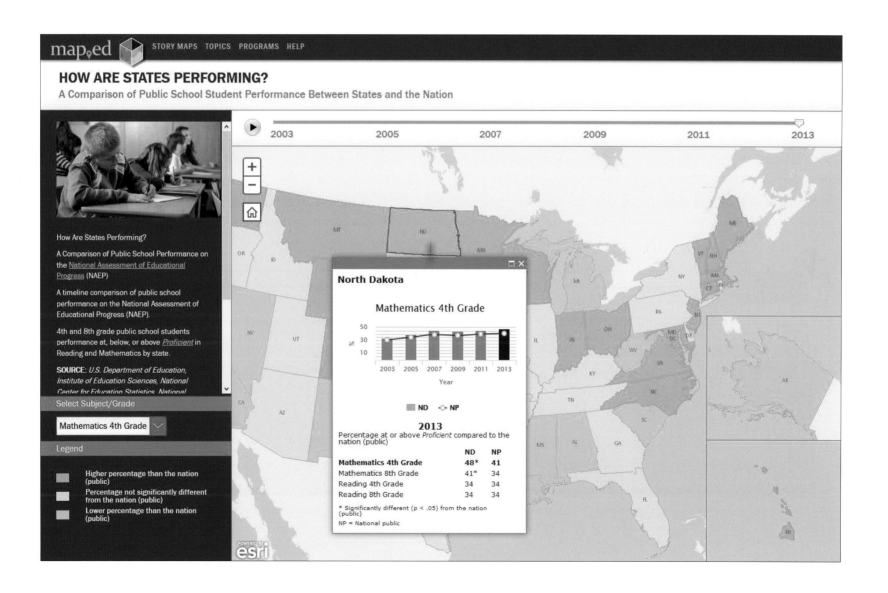

Regional Educational Laboratory and Blue Raster

REL Midwest EdMaps

How has poverty changed in the Midwest region since 2000? The story map features a timeline showing that the number of districts with high levels of students eligible for free and reduced-price lunch has increased considerably over the decade. EdMaps helps education stakeholders think spatially about education issues affecting their communities.

 EdMaps

Questions or comments?
Contact:

Jason Narlock, Ph.D., Researcher
REL Midwest at American
Institutes for Research (AIR)
jnarlock@air.org
P: 312-588-7346

FAQs

Regional Priority Story Maps

Create a Custom Map →

REL Midwest EdMaps is a Web-mapping application that displays complex education data geographically as "story maps." EdMaps helps users analyze publicly available school- and district-level data, observe trends over time, and frame conversations on practice and policy.

Read More +

| All Maps | College and Career Readiness | Early Childhood Education | Educator Effectiveness | Low-Performing Schools |

Broadband Speeds and School Locale

Historic District-Level Ethnicity

Priority and Focus Schools

Historic District-Level Poverty

US Department of Agriculture Forest Service, Northern Research Station, Forest Inventory and Analysis

Modeled distributions of twelve tree species in New York

The map details the distribution of twelve tree species in New York via a series of maps, histograms, statistics, and other information, as a resource for instruction of upper elementary to adult students. Trees are often not studied in K-12 classes. Forest inventory data can feel inaccessible. Real-world research can feel remote from classroom activities. Resources are needed for outreach activities and attractive maps invite engagement. More educators will have access to tree species information for instruction. More people will be aware of tree species in their area and how to access available data. With this map we are able to reach audiences that might not otherwise be able to use forest information in their work and education.

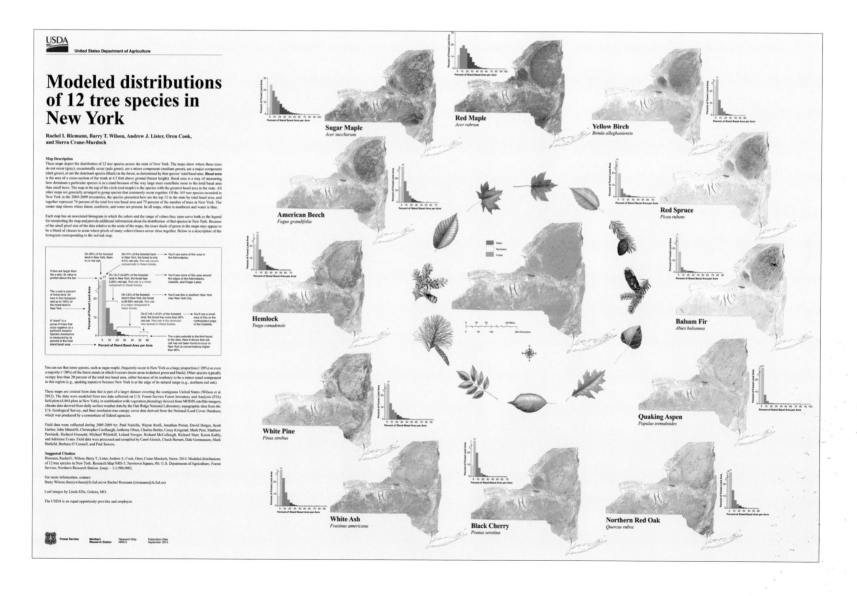

EMPOWERING
HUMANITARIAN
EFFORTS

"By using GIS, project managers can quickly demystify complex problems and assist the decisions that aid and development organizations make every day."

Dustin Horner, director of engagement and partnerships, Development Gateway

Empowering Humanitarian Efforts

Government agencies, international organizations, and non-profits respond to crisis around the world to serve the world's most vulnerable populations. The situations are often dynamic and data is commonly limited or outdated. To overcome these challenges, organizations use GIS in innovative ways to publish open data and strengthen partnerships so they have the best data to inform critical decisions.

Organizations like the World Bank, United Nations, and national governments use GIS to deliver current, robust data that enables smart communities that can act quickly and confidently. They also use GIS to visualize resources, areas of need, and operations for a clear picture of response activities and progress toward sustainable development goals. These efforts are leading the way toward a more sustainable and brighter future.

The World Bank

GeoWB for Data Collection

The World Bank is an international financial institution that provides loans to developing countries for capital programs. The World Bank's internal ArcGIS Portal is called the GeoWB. This application allows World Bank staff to explore and combine GIS from numerous public sources with project specific data in ways not previously possible. The tool allows all World Bank staff the opportunity to investigate data in a novel way, and exposes staff to new, spatially-focused ideas and ways of thinking.

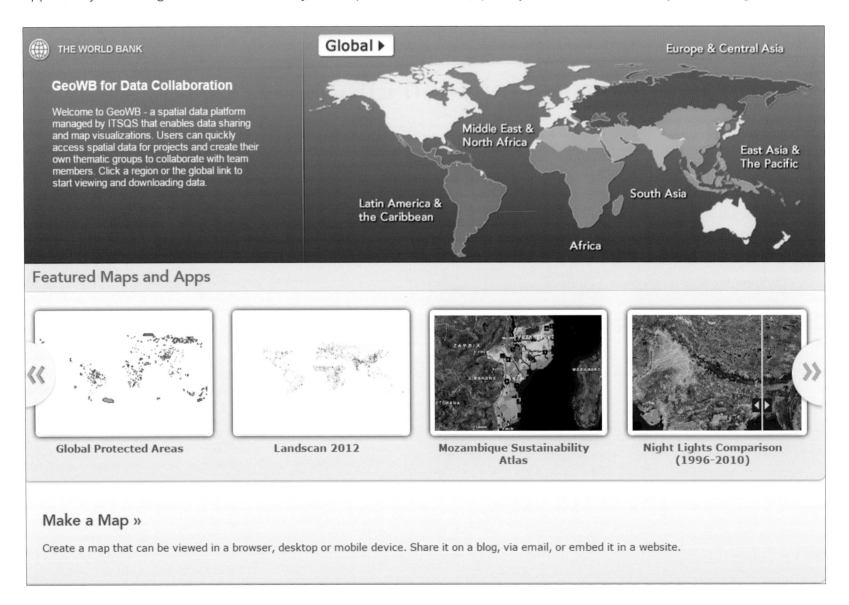

Mapping Poverty in Bolivia

This World Bank map visualizes the number of donor-financed activities in Bolivia per municipality overlaying the incidence of extreme poverty in 2001.

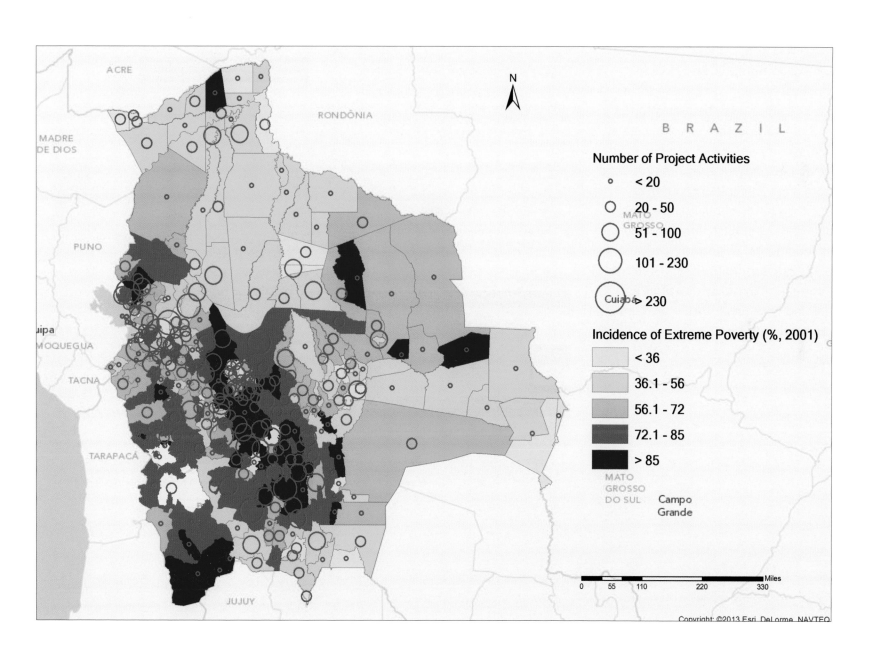

US Agency for International Development (USAID) and Blue Raster

USAID Spatial Data Repository

The Spatial Data Repository (SDR) is an open data tool that provides the global community with new ways of visualizing demographic and health data. This application supports The DHS (Demographic and Health Surveys) program, which assists in the collection and dissemination of global health data. Users can create maps, compare survey regions over time, and customize visuals with data from more than ninety countries and thousands of health indicators and demographic characteristics, including age, education, and wealth. SDR is increasing the use of current information in policy decisions and helping improve accurate program monitoring.

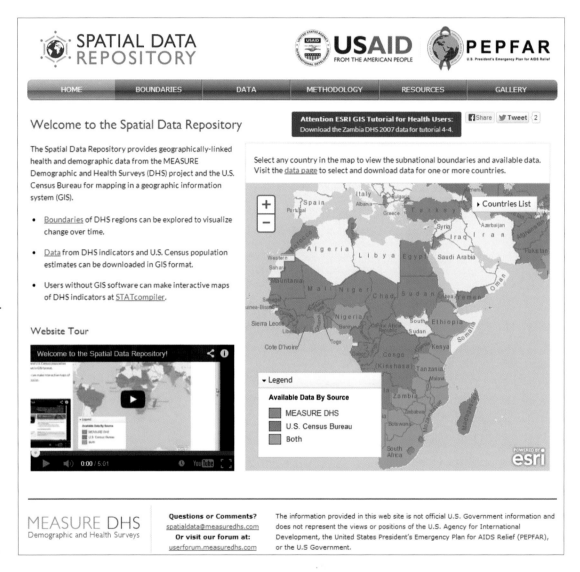

United Nations Office for the Coordination of Humanitarian Affairs (OCHA)

Natural Disaster Activity in the Asia-Pacific Region

This map was intended to raise awareness on the nature of humanitarian work in the Asia-Pacific region. Specifically, the map tries to provide a rough overview of 2012. Humanitarian work in the Asia-Pacific region generally revolves around natural disasters, so this recaps the impact of natural disasters during the year compared to previous years. Also, there is often the impression that as Asia grows and develops there is less and less of a role for the UN and humanitarian partners to play. The map tries to put this notion in context. While there were a number of disasters in which the international humanitarian community played no role, there were nine disasters in which the UN did.

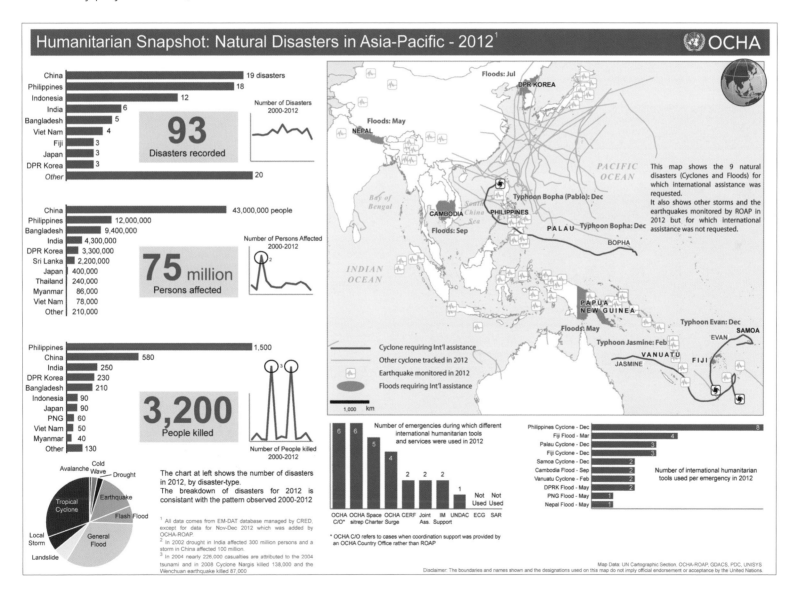

International Committee of the Red Cross (ICRC)

Somalia: Picking Up the Pieces

Thousands of people in Somalia's eastern Puntland region are trying to cope with the aftermath of a devastating cyclone in November 2013 that killed hundreds of people. Also in 2013, farther south in Middle Shabelle, tens of thousands were hit by major flooding by the River Shabelle, especially in and around the town of Jowhar. These story maps show the relief efforts by the International Committee of the Red Cross and Somali Red Crescent.

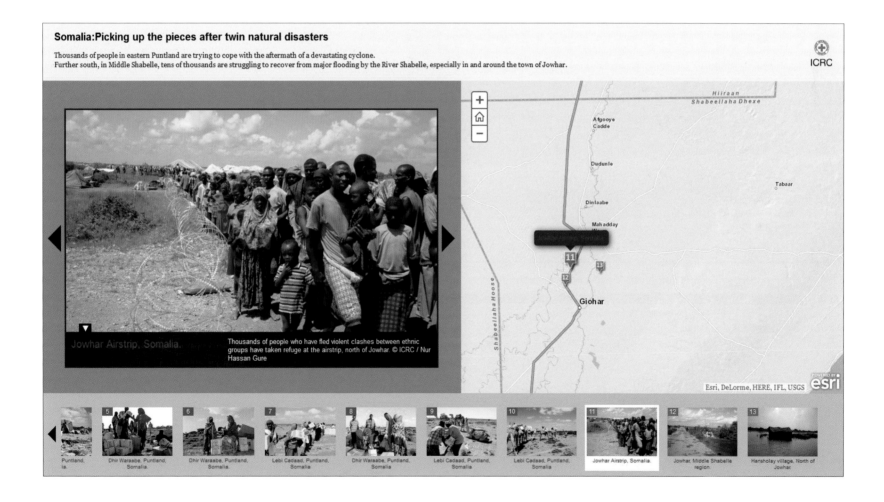

Somalia: Picking up the pieces after twin natural disasters

Thousands of people in eastern Puntland are trying to cope with the aftermath of a devastating cyclone.
Further south, in Middle Shabelle, tens of thousands are struggling to recover from major flooding by the River Shabelle, especially in and around the town of Jowhar.

Jowhar Airstrip, Somalia.

Thousands of people who have fled violent clashes between ethnic groups have taken refuge at the airstrip, north of Jowhar. © ICRC / Nur Hassan Gure

Esri, DeLorme, HERE, IFL, USGS

World Vision International

World Vision International Typhoon Haiyan Response

World Vision is a global Christian relief, development and advocacy organization dedicated to working with children, families, and communities to overcome poverty and injustice. These maps were used to assess World Vision's crisis response in November 2013 to Super Typhoon Haiyan in the Philippines, one of the strongest and most devastating tropical storms ever. World Vision mobilized resources to assist 1.2 million people (240,000 families) with food, nonfood items, hygiene kits, emergency shelter, and protection, especially for children and women.

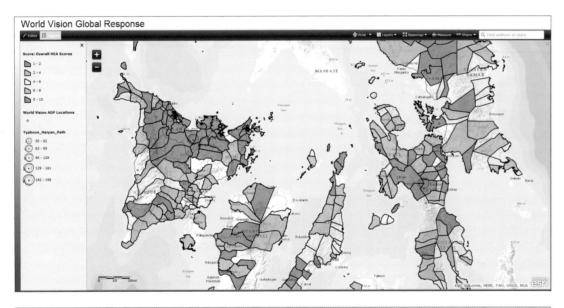

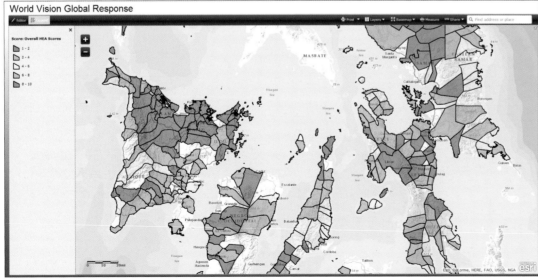

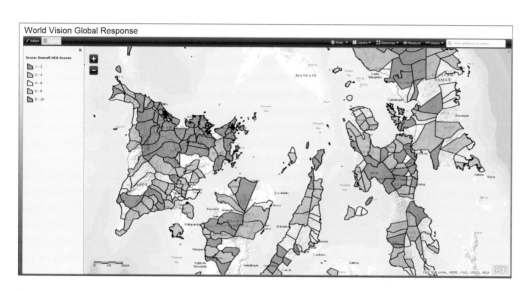

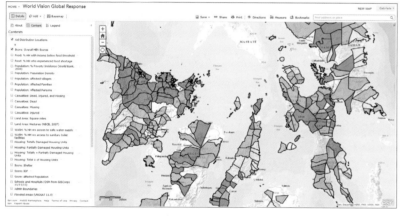

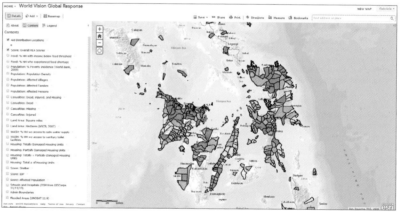

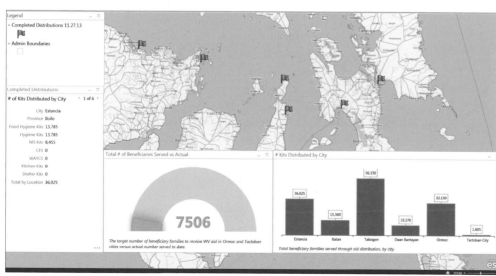

World Resources Institute (WRI) and Blue Raster

WRI Rights to Resources

The World Resources Institute's Rights to Resources tool visualizes resource rights in Africa and allows users to compare findings across countries and resources: water, trees, wildlife, minerals and petroleum. The data displayed in the tool is the outcome of WRI's systematic review of the national framework laws across the five resources. The results of which confirm long suspected notions that across sub-Saharan Africa, few national laws provide communities with strong, secure rights to the resources on their land and that rights to many high-value natural resources are held by the state. The new Rights to Resources application endeavors to strengthen and secure resource rights for communities in sub-Saharan Africa while incentivizing sustainable management of local resources.

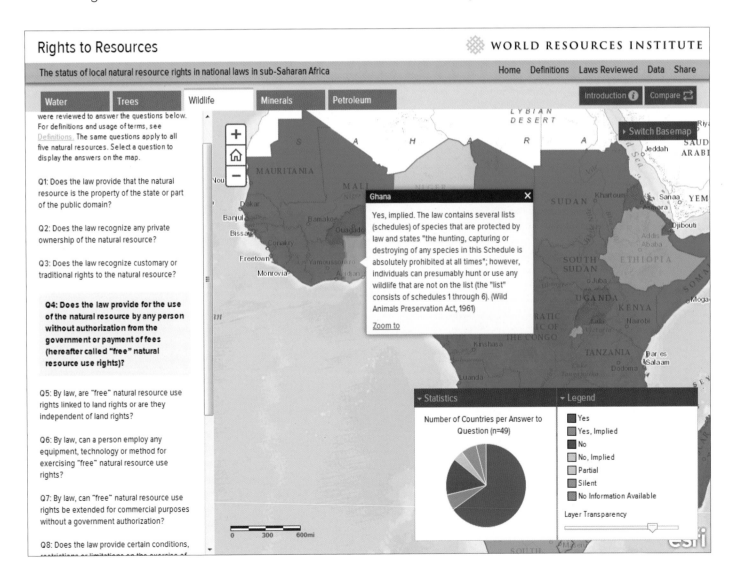

US Agency for International Development (USAID) and Blue Raster

South Sudan: Conflict Causes Major Displacement and Destruction of Markets

How does the US Agency for International Development (USAID) prepare for famine assistance to the world's neediest? One success story is the Famine Early Warning Systems Network (FEWS NET). Two side-by-side comparative maps of food-security outlooks are produced for every region, and the countries are colored according to their food security, highlighting areas of Stress, Crisis, Emergency and Famine. FEWS NET is a collaboration of international, regional, and national partners to provide timely and rigorous early-warning and vulnerability information on emerging and evolving food security issues. Once identified, FEWS NET uses a suite of communications and decision-support products to help decision makers act to mitigate food insecurity. This map focuses on the conflict in South Sudan.

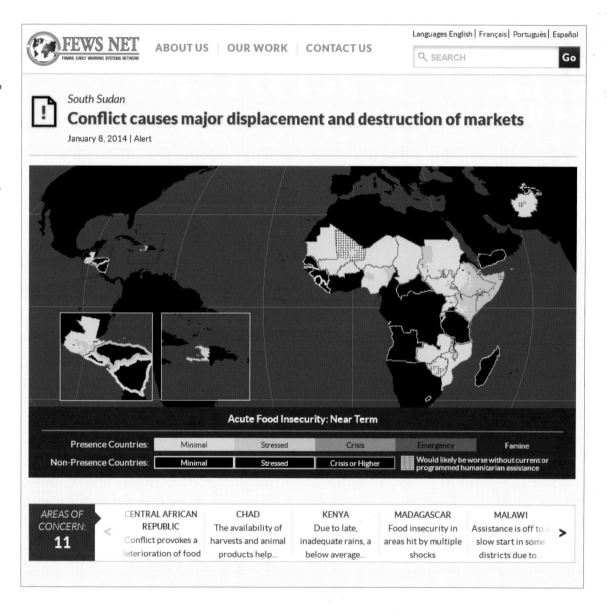

DESIGNING SMARTER INFRASTRUCTURE

Designing Smarter Infrastructure

We often take the road and rail networks we rely on for granted. This transportation infrastructure is a critical part of economic growth, the movement of goods and supplies, and the travel of people from place to place. If the bridges, highways, and rail lines aren't maintained, we risk jeopardizing supply chains, development, and mobility.

In our ever-changing world there is a growing need for movability while simultaneously addressing the issue of aging infrastructure. To respond to these concerns, governments and transportation agencies work to build smarter infrastructure. Smart infrastructure is founded in the idea of strategic investments, longer asset lifecycles, increased safety, and proactive maintenance.

Smart, sustainable infrastructure begins with GIS. Visualization and analysis are key components of successful infrastructure management and execution. GIS integrates the planning, design, construction, and maintenance processes, which allows for more strategic and informed decisions and investments. With analytical capabilities, GIS also helps maximize infrastructure's performance while saving time and money.

Transportation organizations around the world do amazing work to keep nations strong and moving. Their support of smart infrastructure leads to smarter communities and nations that support efficiency, development, and economic prosperity.

US Department of Transportation (DOT)

Potential Sea Level Rises Impact on Transportation Infrastructure and Systems in Mobile, Alabama

The map shows the areas and transportation infrastructure at risk around Mobile, Alabama due to potential sea level rise due to climate change. The map is intended to enhance regional decision makers' ability to understand potential impacts on specific critical components of infrastructure and to evaluate adaptation options.

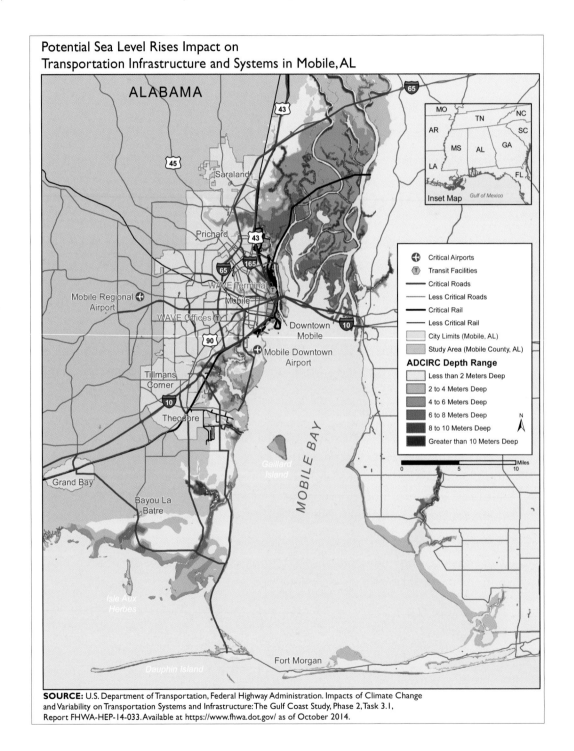

Potential Sea Level Rises Impact on Transportation Infrastructure and Systems in Mobile, AL

SOURCE: U.S. Department of Transportation, Federal Highway Administration. Impacts of Climate Change and Variability on Transportation Systems and Infrastructure: The Gulf Coast Study, Phase 2, Task 3.1, Report FHWA-HEP-14-033. Available at https://www.fhwa.dot.gov/ as of October 2014.

Top Twenty five Busiest Amtrak Stations: FY 2012

Amtrak operates a nationwide rail network, serving more than 500 destinations in forty-six states and three Canadian provinces on more than 21,200 miles of routes. Even with a network this big, some stations welcome more passengers than others. The map shows the top twenty-five Amtrak stations in FY 2012, which are highlighted by total ridership. The map recognizes the locations of the busiest Amtrak stations and where they are relative to one another.

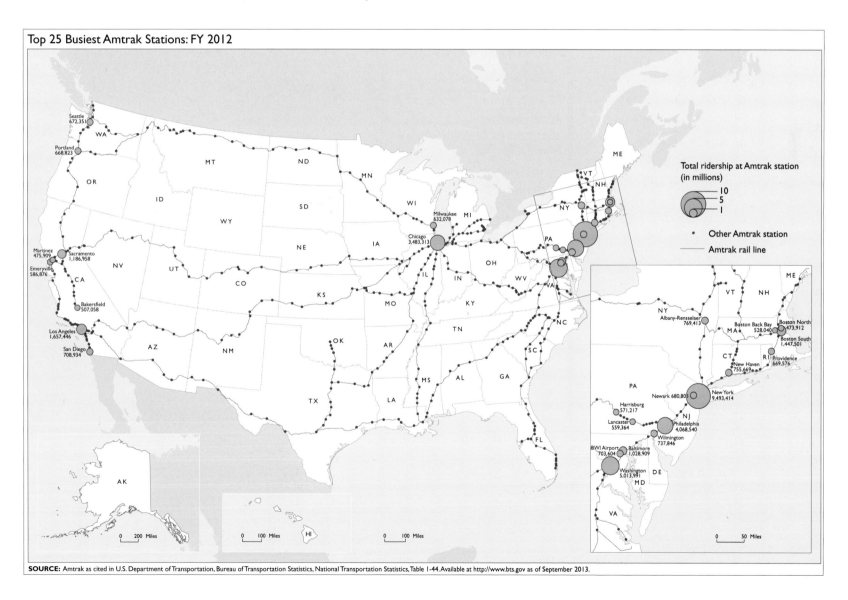

Top 25 Busiest Amtrak Stations: FY 2012

SOURCE: Amtrak as cited in U.S. Department of Transportation, Bureau of Transportation Statistics, National Transportation Statistics, Table 1-44. Available at http://www.bts.gov as of September 2013.

US Department of Transportation, Federal Aviation Administration (FAA)

FAA Infrastructure Map

The FAA Unstaffed Infrastructure Sustainment (UIS) Office manages unstaffed infrastructure that houses, protects, and supports National Airspace System (NAS) communications, navigation, surveillance, weather, and other air traffic control equipment. UIS infrastructure is composed of more than 23,500 facilities supporting seventy-eight equipment and service facility types. The UIS office has developed a GIS system to assist in the management of FAA's UIS infrastructure. UIS GIS is a web-based mapping application designed to allow users to easily query and analyze numerous data sources related to FAA facilities and then visualize the results on a map.

FAA Obstruction Surface Penetration App

This is an application to perform a risk based obstruction surface penetration analysis and intuitively visualize results.

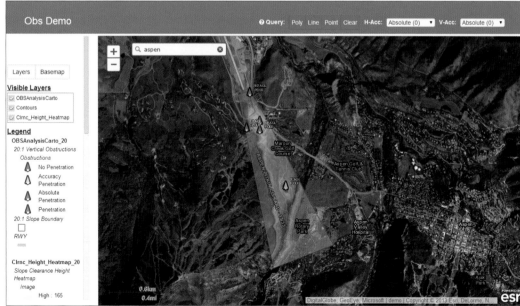

Metron Aviation, Inc.

Mobile App for Unmanned Aerial Systems Flight Planning

Unmanned Aircraft Systems (UAS) operate aircraft with no human pilot on board. Either the aircraft is programmed or the pilot is remote (ground-based). UAS are ideal for operating missions that are too long in duration (e.g. 36 hours) or too hazardous for an onboard pilot. Viable applications include military missions, law enforcement, border patrol, weather data collection, telecommunications, land use imaging, and cargo transport. NASA and other organizations have invested heavily in UAS research. Since today's national airspace system is designed primarily for manned aircraft, integration of UAS into today is an ongoing challenge. Metron Aviation researches and addresses issues associated with UAS integration, such as conflict detection and resolution, control and communication mechanisms, new policies and procedures, and security measures.

National Aeronautics and Space Administration (NASA)

The Stennis Institutional Real Estate Editor

The Stennis Institutional Real Estate Editor (SIREE) allows users to edit information associated with floor plan features for over 200 buildings at the John C. Stennis Space Center, a NASA rocket-testing facility in Mississippi. Users can also locate site employees, edit groups of rooms, and view associated reporting information used for billing.

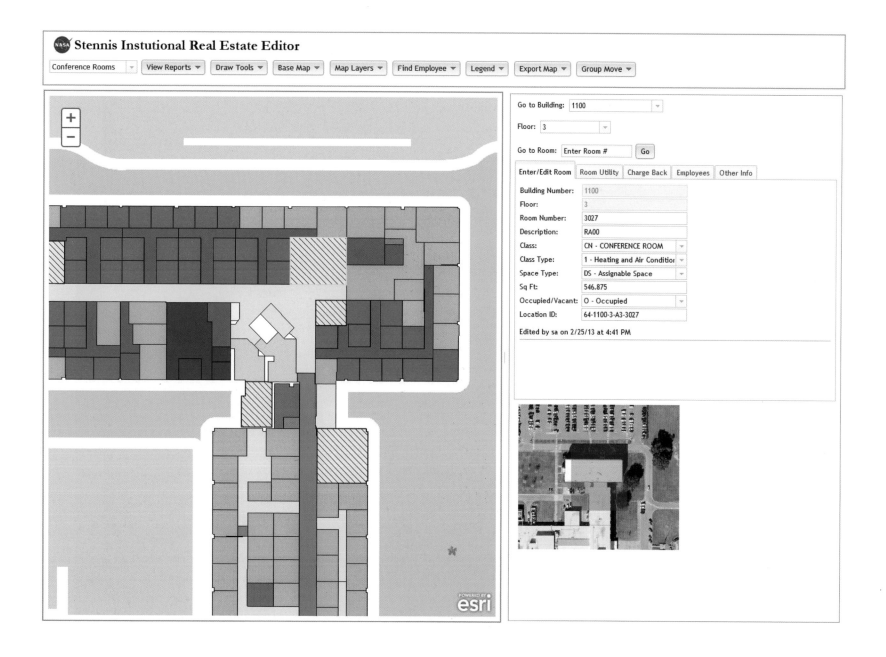

NASA-wide Institutional GIS Portal

The NASA-wide Institutional GIS Portal hosts spatial datasets for ten NASA centers and four component facilities. The portal allows users to view and query agency property information across the nation. A new feature of the portal enables users to create custom calculations based on the real property data and view the custom scenarios on the map. The portal enhanced situational awareness of coastal centers during storm events and saved time by integrating databases and tabular information accessible from desktop computers.

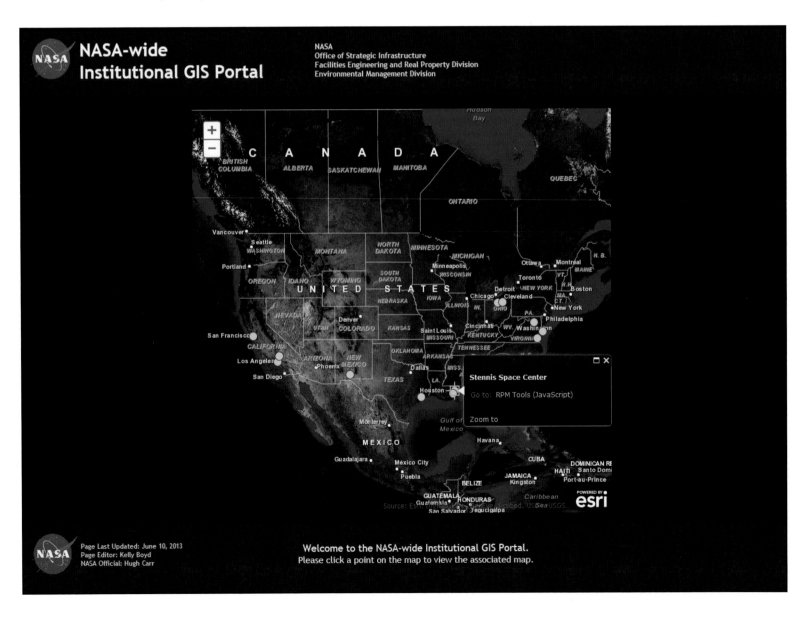

US Postal Service (USPS)

LA Same-Day Delivery

The US Postal Service tested grocery delivery service in the Los Angeles and San Diego areas, part of the organization's efforts to explore new ways to meet customers' changing needs. The test followed up a two-month operational test in San Francisco. Customers could order groceries and other products through Amazon Fresh and have them delivered by the Postal Service early the following morning. In Los Angeles and San Diego, USPS is delivering Amazon Fresh orders throughout the day. To make the deliveries, the Postal Service is using dynamic routing, a technique that sequences a changing series of delivery points in a logical and efficient order. USPS is testing several other products and services, including a Sunday package delivery program.

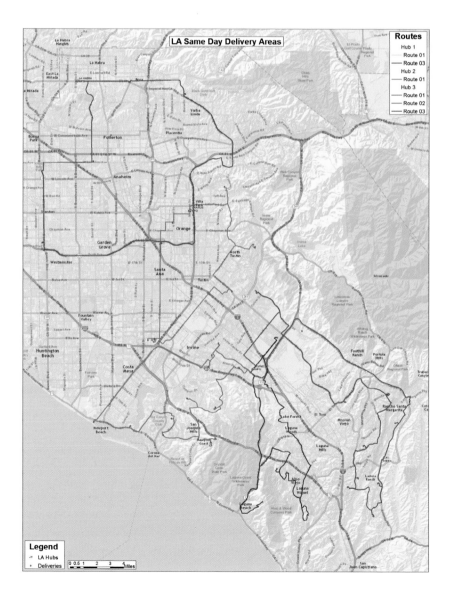

Rural Street Database

The US Postal Service's Rural Street Database allows carriers to identify locations of delivery and attributes of streets and programmatically calculates time and distance for each route. These maps display the setting of address and traffic points along a carrier's line of travel. Each carrier is expected to set all address points including the mail stop (location of where the truck is parked), mail box (location of the mailbox), DDD (location of where a package is delivered) and the door (location of the door at the address). The maps also display and allow carriers to set traffic points along their route. Once all address points and traffic points are set, the carrier's line of travel is completed. Using GIS allows the carriers to identify locations of delivery and attributes of streets as well as calculate time and distance for each route.

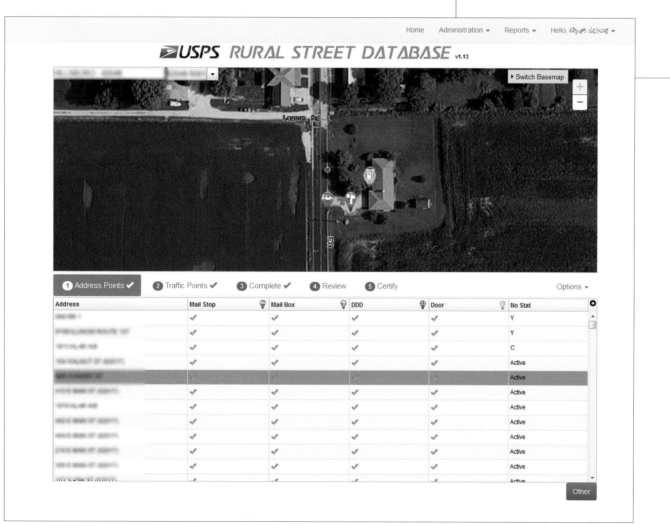

US Postal Service (USPS) Office of the Inspector General

Optimizing Postal Operations Vehicle Maintenance and Repair

Vehicle maintenance and repair is crucial to the everyday operation and mission of the US Postal Service. Postal Vehicle Maintenance Facilities (VMFs) are located throughout the nation and service over 211,000 vehicles, averaging $1.1 billion of yearly maintenance expenses. Each type of postal vehicle is assigned a work credit value based on level of effort required to maintain and service the vehicle. Each VMF has a accumulation of work credits based on the number and types of vehicles they service by postal district. The current distribution of VMF work credits leaves some postal districts with extremely heavy workloads, while other postal districts have significantly lighter. By balancing work credits geographically, new VMF territories are proposed to reduce transportation costs, increase efficiency, and potentially save the US Postal Service millions of dollars in work-hours and labor expenses.

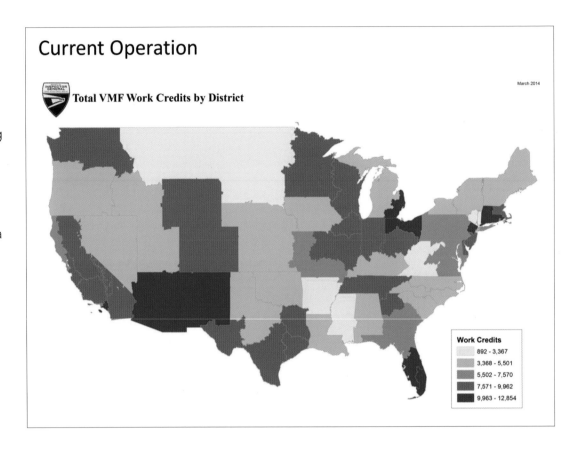

Proposed Optimized Operation

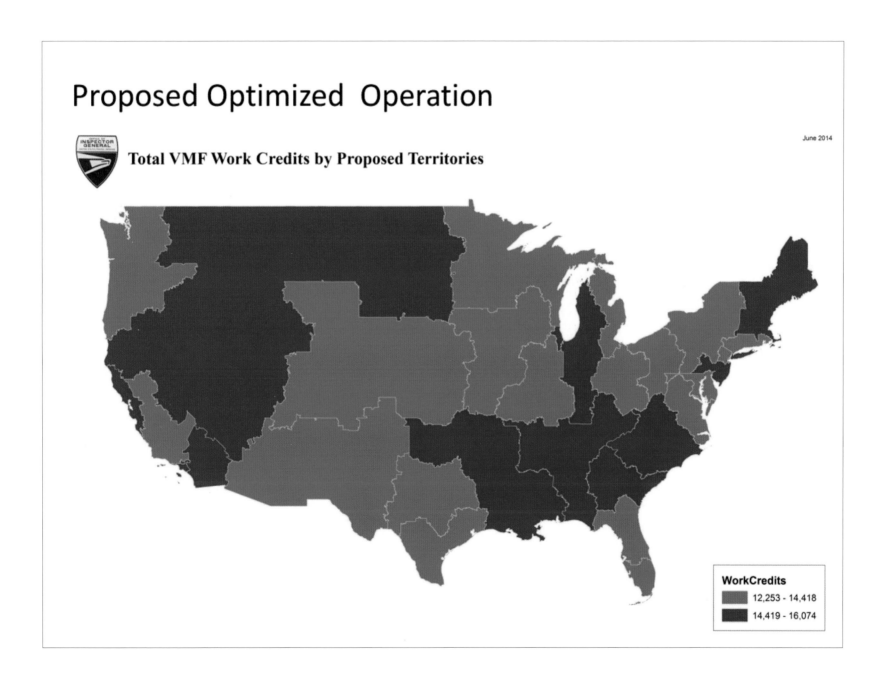

June 2014

Total VMF Work Credits by Proposed Territories

WorkCredits
- 12,253 - 14,418
- 14,419 - 16,074

Delivery Optimization

Since 2010, North Dakota has experienced the nation's fastest-growing population of 7.6 percent compared to national rate of 2.4 percent, mostly due to the oil boom in the Bakken Shale formation. The state's population growth has created a 14 percent increase in the US Postal Service delivery points and a 165 percent in package deliveries. By overlaying the USPS Processing Plants by Size on top of the percentage of population growth for the state of North Dakota, the Office of Inspector General was able to gain insight into the current state of the mail service. The outcome resulted in the North Dakota Postal Service Operations Audit which recommended the installation of additional equipment, consolidation of certain operations, the development of a contingency plan to address staffing, and the modification of the current postal transportation network.

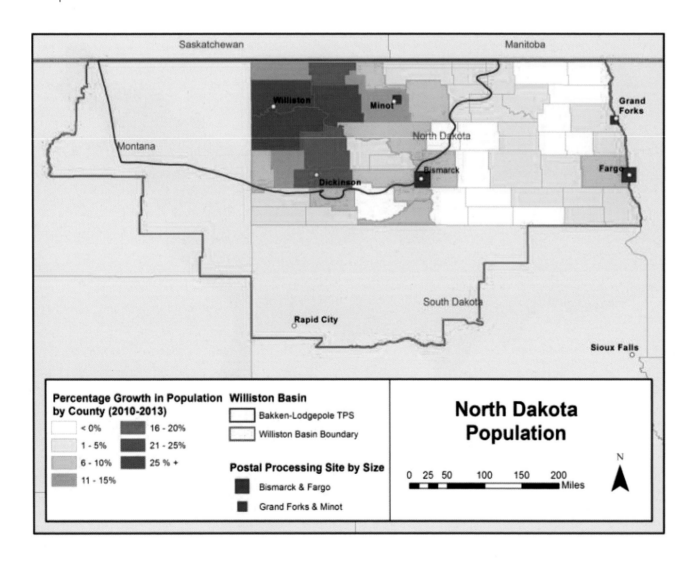

General Services Administration (GSA)

GSA Portal

GSA has over 12,000 employees around the country, most of whom have little or no GIS training and experience but who can benefit from access to GIS tools and spatial analysis. The GSA Portal makes GIS more accessible to the workforce. In addition to its mapping and collaboration capacity, the portal gives employees instant access to existing apps and tools. It's a one-stop shop for access to GIS tools, helping new users identify ways they can start to use GIS to inform and enhance their own business practices.

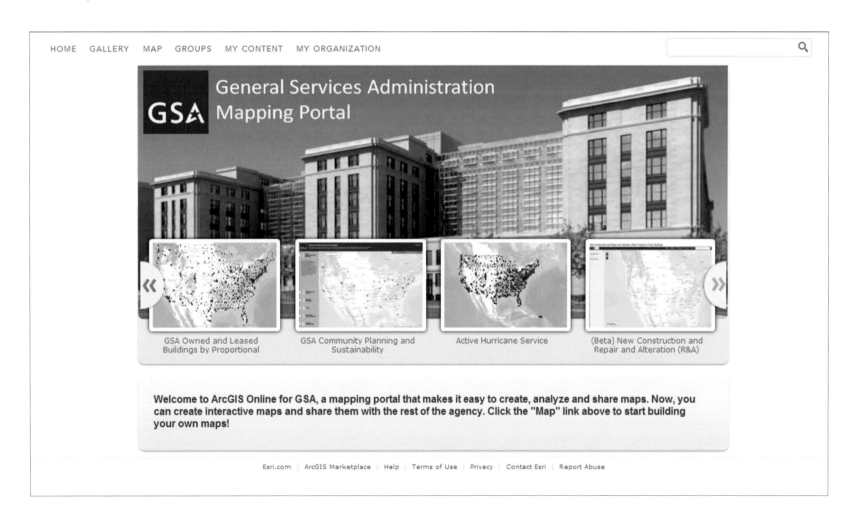

Inventory of Leased Buildings

GSA manages over 9,000 buildings and wanted to share inventory information with the American public. The prior system was outdated and cumbersome. The new Inventory of Owned and Leased Properties Viewer is a custom-built JavaScript tool, available to the public, which displays basic building data for the entire inventory and allows users to view building information and query on certain spatial parameters (e.g. buildings within a particular congressional district). The IOLP viewer supports the president's Open Government initiative, displaying GSA's inventory in a dynamic map and allowing the public to understand GSA's real estate holdings in their community.

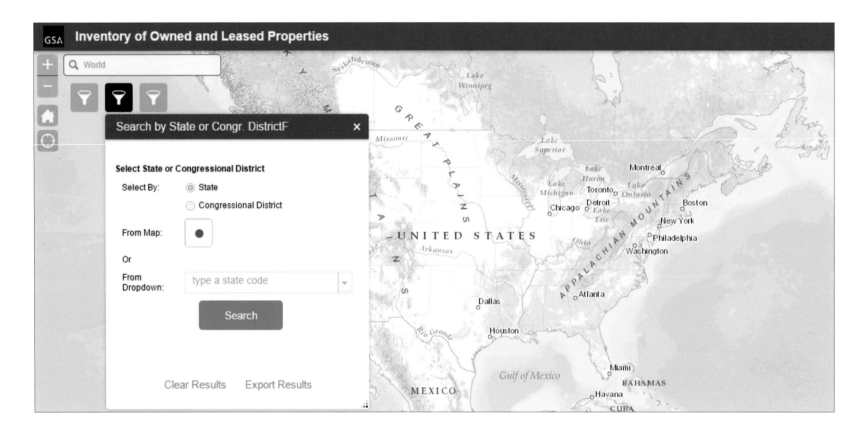

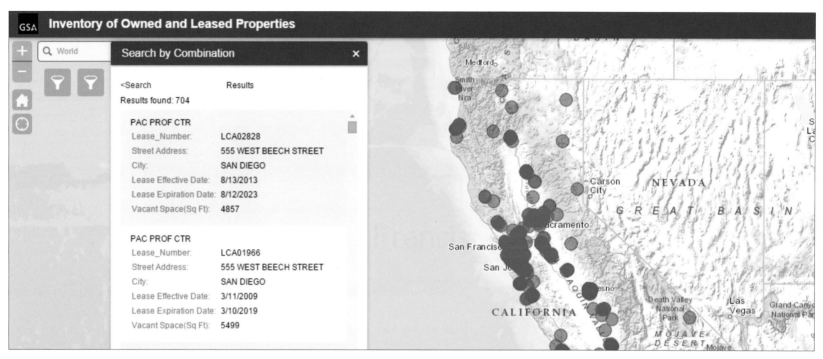

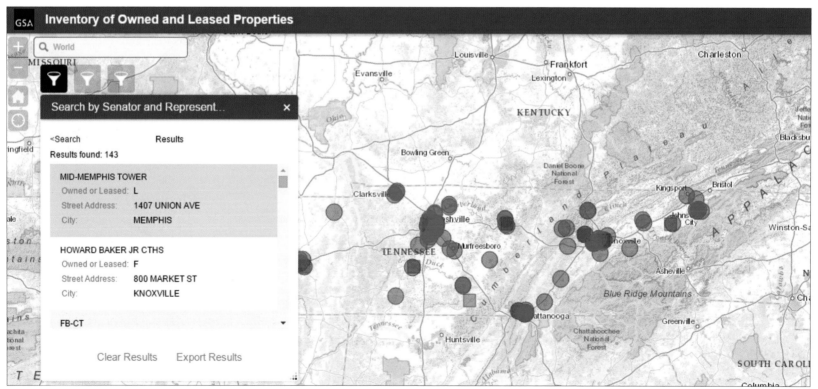

US Navy and Geographic Information Services

Navy Shore Geospatial Energy Module Map Book

Fig 1. A page from the NSGEM mapbook (for the Southwest region). Color-coding illustrates the energy consumption at each lens (Navy, Regional, Installation and Facility). Mapbooks are comprehensive by region and are downloadable and prortable.

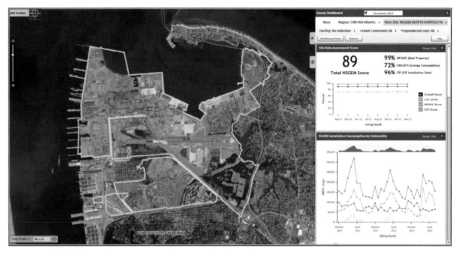

Fig 2. Screenshot of the NSGEM application at the Installation level. As the Navy tries to reduce overall energy usage, NSGEM allows for quick visualization of high consuming facilities. Installation Energy Managers can now target those buildings to meet consumption goals.

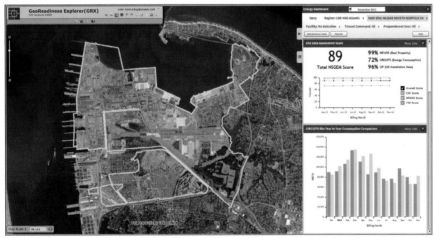

Fig 3. Screenshot of the NSGEM application at the Installation level. As the Navy tries to reduce overall energy usage, NSGEM allows for quick visualization of high consuming facilities. Charts and graphics offer consumption comparison so that Energy Managers can see improvements with each billing month.

Fig 4. A snapshot from the NSGEM mapbook. Color-coding illustrates the energy consumption at each lens (Navy, Regional, Installation and Facility). Mapbooks are comprehensive by region and are downloadable and portable.

Navy Region Southeast Energy Use

This map displays a roll-up of each site's energy use within the Southeast Navy Command Region. The region is headquartered in Jacksonville, Florida and includes installations in 12 states: Kansas, Oklahoma, Texas, Missouri, Arkansas, Louisiana, Tennessee, Mississippi, Alabama, Georgia, Florida, and South Carolina. The map shows energy use over time and provides analysis to drive decision makers' choices

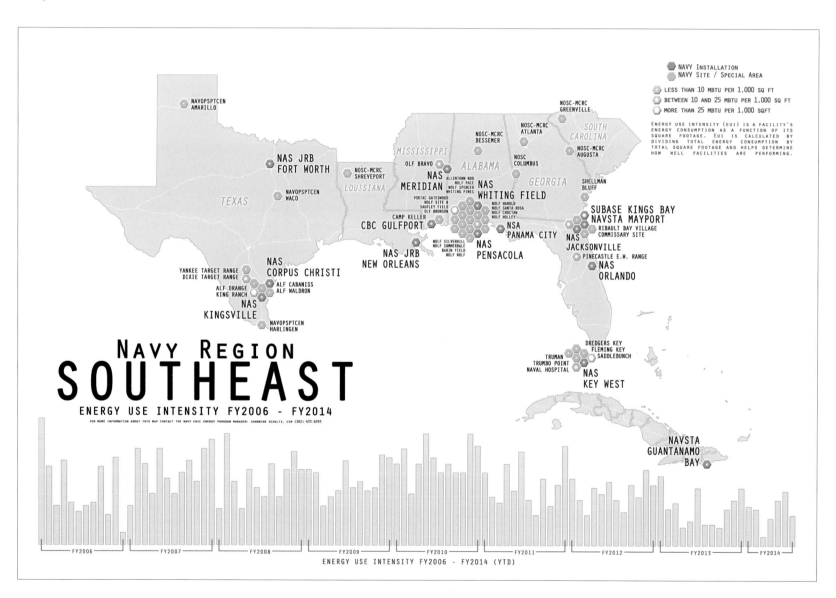

National Oceanic and Atmospheric Administration (NOAA) Coastal Services Center

Exploring Ocean Wind Energy

The offshore energy planning process is complex, and the breadth of data needed to support this process can be overwhelming. This simple story map leads a user through the process of understanding energy potential and defining wind energy planning areas. It also provides examples of data that should be evaluated in order to understand impacts to other ocean uses. This map tells the story of how MarineCadastre.gov data can help with offshore wind planning. MarineCadastre.gov is a joint initiative of the Bureau of Ocean Energy Management and the National Oceanic and Atmospheric Administration, providing authoritative data to meet the needs of the offshore energy and marine planning communities.

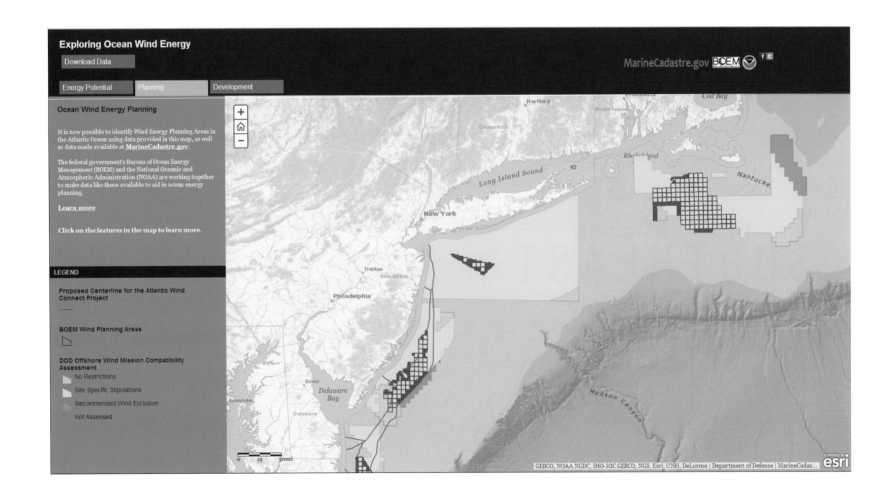

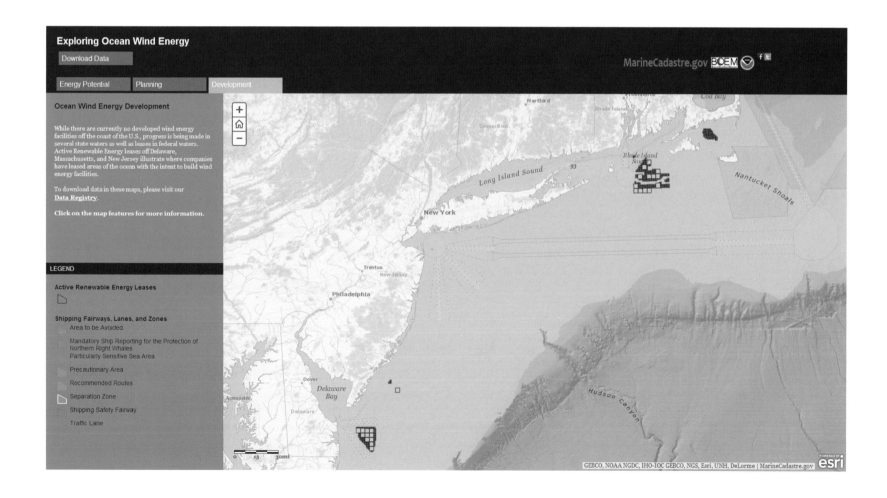

Exploring Ocean Wind Energy

Download Data

MarineCadastre.gov BOEM

Energy Potential | Planning | Development

Ocean Wind Energy Development

While there are currently no developed wind energy facilities off the coast of the U.S., progress is being made in several state waters as well as leases in federal waters. Active Renewable Energy leases off Delaware, Massachusetts, and New Jersey illustrate where companies have leased areas of the ocean with the intent to build wind energy facilities.

To download data in these maps, please visit our **Data Registry**.

Click on the map features for more information.

LEGEND

Active Renewable Energy Leases

Shipping Fairways, Lanes, and Zones

Area to be Avoided

Mandatory Ship Reporting for the Protection of Northern Right Whales Particularly Sensitive Sea Area

Precautionary Area

Recommended Routes

Separation Zone

Shipping Safety Fairway

Traffic Lane

GEBCO, NOAA NGDC, IHO-IOC GEBCO, NGS, Esri, UNH, DeLorme | MarineCadastre.gov

esri

Credits

Architecting a Safe Nation

Ball State University

Atmospheric Rivers and Heavy West Coast Precipitation Events created by Ball State University student Mitchell Pettit; data from National Oceanic and Atmospheric Administration's Monthly US Climate Reports, Federal Emergency Management Agency's Disaster Declarations Summary (1953–present); Ball State University Library GIS Database.

US Air Force and Marstel-Day

The Joint Base Elmendorf-Richardson Mission created by Marstel-Day; data from the Joint Base Elmendorf-Richardson Mission.

US Department of Agriculture (USDA) Forest Service

Kernel Density Wildland Fire Acres Burned Analysis created by USDA Forest Service; data from Short, Karen C. 2015. Spatial Wildfire Occurrence Data for the United States, 1992–2013 [FPA_FOD_20150323]. 3rd Edition. Fort Collins, Colorado: Forest Service Research Data Archive. http://dx.doi.org/10.2737/RDS-2013-0009.3, USDA Forest Service, Francis Marion and Sumter National Forests geospatial program library.

US Army Corps of Engineers (USACE)

Delano, Minnesota, Undocumented Levees Project created by and data from the US Army Corps of Engineers (USACE).

Northern California Regional Intelligence Center (NCRIC)

NCRIC Market Street Security Camera created by Northern California Regional Intelligence Center; data from San Francisco Sheriff's Department based on in-person interviews.

New York City Fire Department (FDNY) and PenBay Solutions

FDNY Super Bowl Event Management created by PenBay Solutions; data from New York City Fire Department (FDNY).

National Cancer Institute (NCI) and Information Management Services

Breast Cancer Mortality Rates created by Information Management Services; data from National Cancer Institute.

US Army Europe, Training Support Activity Europe and Tierra Plan

Army Training Support Activity Web App created by Tierra Plan; data from US Army Europe (USAREUR), Training Support Activity Europe (TSAE).

Virginia Department of Emergency Management

Virginia Department of Emergency Management Webpage created by and data from the Virginia Department of Emergency Management.

US Army Engineer School (USAES)

US Army Engineer School Maps, Jalabad, Afghanistan created by and data from US Army Engineer School (USAES).

Mapping Public Policy

US Senate

US Senator James Risch's Map Gallery created by the office of Senator Risch; data from Esri, DeLorme, HERE, US Geological Survey, National Geospatial Intelligence Agency. Food and Agriculture Organization of the United Nations, National Parks Service, US Environmental Protection Agency, Earthstar Geographics.

US Senator Ron Wyden's Chronic Disease Maps created by the office of Senator Wyden; data from Esri, DeLorme, HERE, US Geological Survey, National Geospatial Intelligence Agency.

US Senator Ron Wyden's Wildfire Maps created by the office of Senator Wyden; data from Esri, DeLorme, HERE, US Geological Survey, National Parks Service, Food and Agriculture Organization of the United Nations.

US House of Representatives

Access Public Transit in California's 41st District created by the office of Representative Mark Takano; data from the Riverside Transit Authority.

US Library of Congress—Congressional Research Service (CRS)

Federal Facilitated Exchanges for the Affordable Care Act Enrollment created by CRS; data from US Department of Health and Human Services; plan selections by ZIP Code in the Health Insurance Marketplace, April 2015; and Esri Data and Maps 2014.

Natural Gas Vehicles created by CRS; data from the US Department of Energy, Alternative Fuels Data Center, and Esri World Street map service..

Islamic State Financing created by CRS; Map boundaries and information generated by Hannah Fischer using Department of State Boundaries (2011); Esri (2013); HIS pipeline, refinery, and electricity plant data (May 2014); Reuters, "Iraq Wheat Silos Held by Islamic State," August 2014; National Geographic, "Survey of Iraq's Archaeological Sites," June 2003; Syria Ministry of Culture, "Archaeological Heritage in Syria During the Crisis, 2011–2013," 2013; Global Reservoir and Dam Database (2010); oil field data provided by the State Department to the Library of Congress (2013) the Economist (2014); and internal expertise.

Credits

Fostering a Healthy Nation

US Food and Drug Administration (FDA)

Application of GIS in Work Planning Process: Service Area Analysis created by US Food and Drug Administration (FDA); data from FDA Office of Regulatory Affairs (ORA), ORA Reporting Analysis and Decision Support System (ORADSS) as of November 2014, OpenStreetMap © OpenStreetMap (and) contributors, CC-BY-SA.

Evaluating the Wastewater Treatment Plant Impact on Shellfish Growing Areas in Alabama's Mobile Bay created by and data from the US Food and Drug Administration.

FIT-Map Mobile App created by and data from the US Food and Drug Administration.

FDA GeoWeb created by and data from the US Food and Drug Administration, Office of Crisis Management

US Environmental Protection Agency (EPA)

Naturally Occurring Anthrax and the Geochemistry of American Soils created by and data from the US Geological Survey.

US National Library of Medicine

Changes for the US National Library of Medicine TOXMAP created by the US National Library of Medicine (NLM); data from NLM, US Environmental Protection Agency.

US Department of State, Humanitarian Information Unit

North Korea: Chronic Food Shortage Coping Mechanisms, Hillside Farming created by and data from the US Department of State.

US Geological Survey

Distribution of Lead (Pb): Soil C Horizon created by and data from the US Geological Survey.

National Cancer Institute (NCI), Centers for Disease Control and Prevention (CDC), and Information Management Services

HPV and Cervical Cancer, Health Information National Trends Survey created by Information Management Services; data from the National Cancer Institute (statecancerprofiles.cancer. gov and hints.cancer.gov)

Centers for Disease Control and Prevention (CDC)

CDC Ebola Response Maps created by and data from Centers for Disease Control and Prevention.

Heart Disease Death Rate created by and data from Centers for Disease Control and Prevention.

US Department of Health and Human Services, Office of Head Start

Children Living in Poverty created by Office of Head Start; data from US Department of Health and Human Services, Esri, DeLorme, Food and Agriculture Organization of the United Nations, US Geological Survey, National Oceanic and Atmospheric Administration, US Environmental Protection Agency.

Supporting a Resilient Environment

National Oceanic and Atmospheric Administration (NOAA)

NOAA OAR PMEL Earth-Ocean Interactions Program created by National Oceanic and Atmospheric Administration (NOAA) Pacific Marine Environmental Laboratory's Earth-Ocean Interactions Program; data from NOAA with bathymetry provided by Monterey Bay Aquarium Research Institute (MBARI), remotely operated vehicle (ROV) imagery and navigation from Woods Hole Oceanographic Institute.

US Army Corps of Engineers (USACE)

Cooper River Critical Wetland Habitat created by and data from US Army Corps of Engineers (USACE).

US Department of Agriculture (USDA)

Northern Minnesota NAIP Imagery created by and data from National Agriculture Imagery Program (NAIP), US Department of Agriculture Farm Service Agency, Aerial Photography Field Office.

United Nations Environment Programme

Sea Ice Concentration and the State of the Polar Bear created by United Nations Environment Programme; data from Atlas of the Cryosphere: Northern Hemisphere; G.W. Johnson, A.G. Gaylord, J.J. Brady, R. Cody, M. Dover, J.C. Franco, D. Garcia-Lavigne, J.C. Gonzalez, W.F. Manley, R. Score, and C.E. Tweedie, 2009, Arctic Research Mapping Application (ARMAP). Englewood, Colorado USA: CH2M HILL Polar Services. Digital Media. http://www.armap.org

US Energy Information Administration (EIA)

EIA Energy Maps created by and data from US Energy Information Administration.

Urban Forestry Administration, District of Columbia Department of Transportation

Urban Forest in Washington, DC created by Urban Forestry Administration, District of Columbia Department of Transportation; data from Esri, DC GIS Open Data.

North Carolina. Bureau of Ocean Energy Management (BOEM) and Geodynamics

Potential Wind Energy Lease Blocks Offshore Cape Fear created by Geodynamics, Bureau of Ocean Energy Management (BOEM); data from BOEM, National Oceanic and Atmospheric Administration (NOAA).

Credits

Optimizing National Mapping and Statistics

US Bureau of Land Management (BLM)

BLM's National Conservation Lands: Protecting Large Landscapes created by US Bureau of Land Management (BLM); data from BLM, US Census Bureau, Esri, Conservation Biology Institute; Esri, DeLorme, General Bathymetric Chart of the Ocean (GEBCO), National Oceanic and Atmospheric Administration (NOAA) National Geophysical Data Center (NGDC)

Public Information Map—Assayii Lake Fire created by US Bureau of Land Management; data from Southwest Area Incident Management Team 3, Navajo Nation Naschitti Chapter, US Geological Survey (USGS), Federal Emergency Management Agency (FEMA).

US Department of Agriculture (USDA) Natural Resources Conservation Service (NRCS)

Characterization of Playa Hydrology on the Southern High Plains created by and data from US Department of Agriculture (USDA) Natural Resources Conservation Service (NRCS)

Appalachian Mountain Club

Land Conservation in Maine: 1994–2014 created by Appalachian Mountain Club; data from Maine Office of GIS, Department of Agriculture, Conservation and Forestry Division of Parks and Public Lands, Land Use Planning Commission, Department of Inland Fisheries and Wildlife, State Planning Office, The Nature Conservancy, New England Forestry Foundation, Maine private land trusts, US Park Service, US Fish and Wildlife, Maine municipal towns, and Appalachian Mountain Club, Geographic Data Technology, Inc., US Geological Survey, US Census Bureau.

US Geological Survey (USGS)

Quaternary Geologic Map of the North-Central Part of the Salinas River Valley and Arroyo Seco created by and data from the US Geological Survey (USGS).

3D Elevation Program: Status of 3DEP Quality Data created by and data from the US Geological Survey (USGS).

US Census Bureau

Census Data Mapper Homeowner Vacancy Rate created by and data from the US Census Bureau.

Census Bureau's TIGER Map Viewer and Web Mapping Services created by and data from the US Census Bureau.

Cabo Verde National Statistics Institute

Census 2010 Project created by and data from Cabo Verde National Statistics Institute.

Dutch Kadaster

Time Travel App created by and data from Dutch Kadaster.

Elevating Education

University of Maryland

Campus Webmap created by the University of Maryland (UMD) Facilities Management; data from UMD Facilities Management, Esri, DeLorme, Food and Agriculture Organization of the United Nations, US Geological Survey, National Oceanic and Atmospheric Administration, US Environmental Protection Agency.

National Center for Education Statistics and Blue Raster

MapED: School Bullying created by Blue Raster; data from US Department of Education Civil Rights Data Collection 2011–2012; Esri

MapED: Public School Student Performance on the National Assessment of Educational Progress created by Blue Raster; data from US Department of Education, Institute of Education Sciences, National Center for Education Statistics, National Assessment of Educational Progress (NAEP), 2003, 2005, 2007, 2009, 2011, and 2013, Mathematics and Reading Assessments; Esri.

Regional Educational Laboratory and Blue Raster

REL Midwest EdMaps created by Blue Raster; data from US Department of Education, Institute of Education Sciences, National Center for Education Statistics, Common Core of Data (CCD), 2000–2012; US Department of Agriculture (2012) National School Lunch Program fact sheet.

US Department of Agriculture (USDA) Forest Service, Northern Research Station, Forest Inventory and Analysis

Modeled Distributions of Twelve Tree Species in New York created by and data from US Department of Agriculture (USDA) Forest Service, Northern Research Station, Forest Inventory and Analysis.

Credits

Empowering Humanitarian Efforts

The World Bank

GeoWB for Data Collection created by The World Bank; data from The World Bank, World Database on Protected Areas, Landscan global population data 2012, Advanced Spaceborne Thermal Emission and Reflection Radiometer (ASTER) Global Digital Elevation Map, National Oceanic and Atmospheric Administration.

Mapping Poverty in Bolivia created by The World Bank; data from The World Bank, Esri, DeLorme, NAVTEQ.

US Agency for International Development (USAID) and Blue Raster

USAID Spatial Data Repository created by Blue Raster; data from Spatial Data Repository, the Demographic and Health Surveys Program, ICF International (spatialdata.dhsprogram.com).

United Nations Office for the Coordination of Humanitarian Affairs (OCHA)

Natural Disaster Activity in the Asia-Pacific Region created by United Nations Office for the Coordination of Humanitarian Affairs (OCHA); data from the Centre for Research on the Epidemiology of Disasters (CRED) International Disaster Database (EM-DAT), UN Cartographic Section, Global Disaster Alert and Coordination System, Pacific Disaster Center, UNISYS.

International Committee of the Red Cross (ICRC)

Somalia: Picking Up the Pieces created by and data from International Committee of the Red Cross (ICRC).

World Vision International

World Vision International Typhoon Haiyan Response created by and data from World Vision International.

World Resources Institute (WRI) and Blue Raster

WRI Rights to Resources created by Blue Raster; data from World Resources Institute.

US Agency for International Development (USAID) and Blue Raster

South Sudan: Conflict Causes Major Displacement and Destruction of Markets created by Blue Raster; data from US Agency for International Development (USAID), Famine Early Warnings Systems Network (FEWS NET)